Batteries for Electric Vehicles

This fundamental guide will teach you the basics of battery design for electric vehicles. Working through this book, you will understand how to optimise battery performance and functionality, whilst minimising cost and maximising durability.

Beginning with the basic concepts of electrochemistry, the author moves on to describe implementation, control, and management of batteries in real vehicles, with respect to the battery materials. The author describes how to select cells and batteries with explanations of the advantages and disadvantages of different battery chemistries, enabling you to put your knowledge into practice and make informed and successful design decisions, with a thorough understanding of the trade-offs involved.

The first of its kind, and written by an industry expert with experience in academia, this is an ideal resource both for students and researchers in the fields of battery research and development, as well as for professionals in the automotive industry extending their interest towards electric vehicles.

Including a foreword by Leif Johansson, Chairman of Telefonaktiebolaget LM Ericsson and AstraZeneca PLC, and former CEO of the Volvo Group.

Helena Berg is the CEO of AB Libergreen, founded by herself in 2012 to advise other companies in the areas of electromobility and batteries. Previously she was the Global Corporate Battery Specialist of the Volvo Group and she also has a Ph.D. in battery materials.

Batteries for Electric Vehicles

Materials and Electrochemistry

HELENA BERG

CAMBRIDGE
UNIVERSITY PRESS

University Printing House, Cambridge CB2 8BS, United Kingdom

One Liberty Plaza, 20th Floor, New York, NY 10006, USA

477 Williamstown Road, Port Melbourne, VIC 3207, Australia

314-321, 3rd Floor, Plot 3, Splendor Forum, Jasola District Centre, New Delhi - 110025, India

103 Penang Road, #05-06/07, Visioncrest Commercial, Singapore 238467

Cambridge University Press is part of the University of Cambridge.

It furthers the University's mission by disseminating knowledge in the pursuit of
education, learning and research at the highest international levels of excellence.

www.cambridge.org
Information on this title: www.cambridge.org/9781107085930

© Helena Berg 2015

First published 2015

A catalogue record for this publication is available from the British Library

Library of Congress Cataloging in Publication data
Berg, Helena.
Batteries for electric vehicles : materials and electrochemistry / Helena Berg.
    pages   cm
Includes bibliographical references.
ISBN 978-1-107-08593-0 (Hardback)
1. Electric vehicles–Batteries.   I. Title.
TL220.B427 2015
629.25´02–dc23   2015006511

ISBN 978-1-107-08593-0 Hardback

Contents

Foreword   *page* ix
Preface   xi

Introduction   1

I Electrochemistry and battery technologies   5

1   The electrochemical cell   7
   1.1   Definitions   8
   1.2   Cell components   8
        1.2.1   Electrodes   9
        1.2.2   Electrolytes   9
        1.2.3   Separators   10
        1.2.4   Current collectors   10
        1.2.5   Casing   11
   1.3   Cell and battery   11
        1.3.1   Half cells   11
        1.3.2   Full cells: monopolar and bipolar   12
        1.3.3   Full cells: three-dimensional   13
        1.3.4   Battery   14
   1.4   Thermodynamics   14
        1.4.1   Chemical and electrochemical potentials   15
        1.4.2   Cell voltage   17
        1.4.3   Temperature   20
   1.5   Electrode and electrolyte processes 21
        1.5.1   Electrode kinetics   22
        1.5.2   Electrode–electrolyte interfaces   23
        1.5.3   Mass transport   24
        1.5.4   Ion transport   25
        1.5.5   Mass transport in solid states   28
        1.5.6   Electrolyte stability   31
   1.6   Practical cell measures   32
        1.6.1   Cell voltage under load   33

	1.6.2	Charge and discharge rates	35
	1.6.3	Capacity	36
	1.6.4	Energy and power	38
	1.6.5	Efficiency	40
1.7	Electrochemical analysis methods		41
	1.7.1	Galvanostatic and potentiostatic cycling	41
	1.7.2	Cyclic voltammetry	42
	1.7.3	Electrochemical impedance spectroscopy	43
	1.7.4	Reference electrode	46

2   Battery technologies for electric vehicles   47
2.1	Lead-acid batteries		48
	2.1.1	Basics	48
	2.1.2	Lead-acid concepts	51
2.2	Nickel metal-hydride batteries		52
	2.2.1	Basics	53
	2.2.2	NiMH battery materials	55
2.3	Lithium batteries		57
	2.3.1	Lithium metal	58
	2.3.2	Li-ion and Li-ion polymer	59
	2.3.3	Lithium-oxygen	59
	2.3.4	Lithium-sulphur	60
2.4	Electrochemical double-layer capacitors		62
	2.4.1	Capacitor materials	65
	2.4.2	High-energy capacitors	65
2.5	Other battery technologies		66
	2.5.1	High-temperature molten-salt batteries	66
	2.5.2	Nickel zinc batteries	68
	2.5.3	Zinc-air batteries	69
	2.5.4	Metal-ion batteries	70
	2.5.5	Redox flow batteries	72
2.6	Fuel cells		74
	2.6.1	Polymer electrolyte membrane fuel cells	75
	2.6.2	PEMFC usage	78

II Li-ion battery technology – materials and cell design   81

3   Lithium battery materials   83
3.1	Negative electrode materials		86
	3.1.1	The solid electrolyte interphase	87
	3.1.2	Metallic lithium	89
	3.1.3	Carbons	91

	3.1.4	Alloys	96
	3.1.5	Oxides	98
3.2	Positive electrode materials		100
	3.2.1	Layered materials	102
	3.2.2	The cubic spinel $LiMn_2O_4$	105
	3.2.3	Olivine $LiFePO_4$	109
	3.2.4	Other materials	111
	3.2.5	Mixed electrode concepts	113
3.3	Electrolytes and separators		114
	3.3.1	Liquid electrolytes	115
	3.3.2	Separators	120
	3.3.3	Polymer-based electrolytes	123
	3.3.4	Ionic liquids as electrolytes	124

4   Cell design   126

4.1	Composite electrodes		126
4.2	Energy and power-optimised electrodes		129
4.3	Energy and power-optimised cells		130
	4.3.1	Cell balancing	130
	4.3.2	Energy and power relationship	131
	4.3.3	Example: energy and power-optimised cells	132
4.4	Cell format and design		134
	4.4.1	Cylindrical cells	134
	4.4.2	Prismatic cells	135
	4.4.3	Pouch cells	135
	4.4.4	Cell safety devices	137
4.5	Production processes		137
	4.5.1	Safety and reliability	139

III Battery usage in electric vehicles   141

5   Vehicle requirements and battery design   143

5.1	Vehicle types and requirements		143
	5.1.1	Vehicle types	143
	5.1.2	Usage conditions	146
	5.1.3	Energy and power requirements	147
5.2	Battery design		152
	5.2.1	General design criteria	154
	5.2.2	Cell selection	156
	5.2.3	Additional battery components	163
	5.2.4	Design impact on reliability and safety	166

6   Battery control and management   168

6.1   Battery management system   168

  6.1.1   Charge and discharge control and methods   171

  6.1.2   Thermal control and management   174

  6.1.3   Battery monitoring   178

6.2   State functions   179

  6.2.1   State of charge   180

  6.2.2   State of health   189

  6.2.3   State of function   192

7   Battery usage and degradation   194

7.1   Degradation basics and mechanisms   195

  7.1.1   Examples: origins of capacity fade   199

  7.1.2   Accelerated degradation   201

7.2   Degradation of Li-ion cells   208

  7.2.1   General degradation categories   209

  7.2.2   Degradation of active materials   212

  7.2.3   Degradation of electrolytes   219

7.3   Degradation analysis methods   221

  7.3.1   Galvanostatic cycling   222

  7.3.2   Electrochemical impedance spectroscopy   223

  7.3.3   Incremental capacity   224

  7.3.4   Differential voltage   226

  7.3.5   Half cell   227

  7.3.6   Post-mortem   228

Glossary   230

Further reading   234

Index   235

Foreword

Already in the late 1970s when I was the Managing Director of *Husqvarna Motor-cycles*, I was involved in a project where we tried to build useful and light electric scooters. We were forced to give up. The batteries of those days were simply insufficient regarding energy storage. In addition, there were few control components that worked at high enough powers.

Today, 30 years later, we see the first generation of electric vehicles – cars and scooters, as well as city buses – emerge. This is made possible through new types of batteries available in configurations that actually work at high-power outputs and relatively large amounts of stored energy. Today there are also computers capable of monitoring the batteries and there are high-power electronic components based on semiconductors. Altogether this provides the opportunity to construct systems suitable for vehicles. As CEO of the *Volvo Group*, I was happy and proud of the projects emerging with the electrification and hybrid electrification of vehicles during the first decade of the twenty-first century, and with which Helena Berg, among others, was working.

The task is bigger, though, than only supplying vehicles with well-functioning battery packs. As human beings, we are identifying increasing demands on mobility in our everyday life. This implies a desire to make extensive use of mobile devices such as cameras, smart telephones, tablets, media players, and in the future a vast number of products we cannot even imagine today.

'The internet of things' will result in many billions of products needing to communicate with one another in order to establish a society as efficient and accessible as we all wish. All these products will need an energy source most likely a battery. And when building the future electric power supply and distribution system – 'The smart grid' – we will need load levelling and energy storage.

For all this, batteries and battery technologies are needed. We need to deepen our understanding of today's batteries and to better assess what we can expect of batteries in the future.

The knowledge of batteries, battery configurations, and their control has become strategic knowledge that many people need to assimilate. This obviously applies to all categories of product developers and the direct design and construction work, but also to the leaders of such development. I would also argue that it would be beneficial if interested political leaders, developers of society, and decision makers could better understand the possibilities of the technology in such an important field.

Helena Berg has written a book about all of this. She has a profound technical background in addition to a thorough experience of applications in real situations.

Gothenburg,
January 2015

Leif Johansson
Chairman of Telefonaktiebolaget LM Ericsson and AstraZeneca PLC

Preface

When I started to work with batteries 20 years ago, Li-ion cells had been introduced to the market a few years before and everyone was talking about the battery revolution – the electric car will finally become true. Since then we have seen the Li-ion batteries come to totally dominate the consumer electronics market and now starting their journey to become the source of electricity for electric vehicles. Today most vehicle manufacturers are promoting electric vehicles and large electromobility programmes exist among government bodies, universities, and companies around the world as crucial steps towards a sustainable world in terms of meeting the serious threats to our societies such as depletion of oil reserves and climate change.

The key for this to ultimately succeed is knowledge of the battery itself and how to design a battery with optimal performance and functionality at a low cost and with long durability. Trying to design a battery without proper knowledge about the materials used and electrochemistry basics sooner or later ends up in a non-optimal design in terms of cost, performance, or durability. Inside the battery it is the cell chemistry that sets the fundamental limitations and hence, in the long run, also the performance of the vehicle.

This is the book I would have liked to be able to hand out to my co-workers and managers during my years in the automotive industry. This book explains the fundamentals behind why a battery has to be handled according to specific constraints and how it should be matched with the type of vehicle; most of all this book should help design teams to talk the same 'battery language' and thus enable greater battery research.

During my winding road towards a finalised book, I have had the opportunity to work and discuss batteries and electric vehicles with Anette Häger, Erlendur Jónsson, Hanna Bryngelsson, Henrik Engdahl, Jenny Ring, Leif Johansson, Niklas Thulin, Patrik Johansson, Patrik Persson, and Mario Wachtler – all are gratefully acknowledged. A special thanks goes to the professional editorial team at Cambridge University Press who believed in the scope of the book from the very first day.

Helena Berg

Amiens
January 2015

Introduction

Energy efficiency, climate change, and sustainability □ our society is facing a number of major challenges, global as well as local. Our choices of vehicles and transportation solutions in general are often discussed in this context. The global challenges of the vehicle sector are mainly related to fuel consumption as this is directly linked to CO_2 emissions. Politically oil dependency and resources contribute to the agenda. On a local and regional level, the challenges are more related to connectivity and emissions – the transportation of people and goods to, and in, our cities should be as smooth as possible. Noise is one type of emission greatly affecting life in the cities, triggering a desire for 'silent' vehicles (without, of course, reducing the safety of pedestrians). Different kinds of electric vehicles can clearly contribute to solve these issues and help us to reach a more sustainable future for our modes of transport.

There are mainly three kinds of electric vehicles: hybrids, all-electric, and plug-in hybrids. The all-electric vehicle relies solely on electricity and the other two utilise both electricity and an internal combustion engine. Which type of vehicle is preferable depends on several factors, e.g. purposes, manufacturing strategies, cost, and market.

How far can I drive my car with a partially charged battery? How much will the fuel economy be improved? Questions like these rely on the design of the electric vehicle and foremost on the utilisation of the battery. The different electric vehicles have their own specific requirements for energy and power, the challenge being to find the most optimal design in terms of energy consumption and driveability from a sustainability perspective. The energy demands often translate to all-electric driving range, and the power demands to acceleration and driveability.

The enabler of the electric vehicle is thus really the battery – providing the electric energy. Obtaining the optimal energy and power is an art relying on knowledge both of the vehicle and battery design. The energy and power demands can indeed be fulfilled by a number of battery technologies, but no single technology can fulfil them all perfectly as they have different energy and power characteristics. Furthermore, the materials used internally in the batteries enable functionality and determine the performance limitations. The energy density of a battery fundamentally originates from the chemistry of the materials used, while the power capability mainly originates from materials physics and production constraints.

Electric vehicles often require very dynamic usage of the battery to be possible, i.e. rapid charge/discharge often involving high currents, originating in driving patterns of braking and acceleration. These dynamics in combination with operational conditions such as temperature affect battery performance and may result in reduced battery durability. Moreover, battery performance is very much path dependent, i.e. usage history will affect future capacity and power. Therefore, a reliable and robust battery management system is required to keep battery usage within preferred limitations at all times.

The battery management system as such relies to a great extent on models, empirical or physical, which in turn originate from electrochemistry and battery materials. During vehicle operation, the battery management system has to predict the status of the battery to optimise its usage. The predictions are based on the models and on actual measurements of battery parameters such as current, voltage, and temperature. In order to secure durability, the battery is rarely fully discharged, and the discharge actually allowed is highly dependent on the type of vehicle, battery design, and the materials of the battery.

The purpose of this book is to present the fundamentals of battery electrochemistry and the materials involved, from the electric vehicle perspective. This way an understanding can be obtained of short-term, as well as long-term, battery life and usage. The goal of electric vehicle battery designers is a long-lasting battery of low cost, and therefore battery degradation processes are treated in detail, including the role of usage.

Indeed, when designing a battery a number of parameters must be balanced: e.g. energy and power demand, thermal specifications, mechanical stress, size and weight, and vehicle integration. When finally selecting a battery, additional factors such as availability, production capabilities, manufacturing qualities and warranties, and cost must be taken into account. How different conditions of battery usage as well as storage affect durability and performance is in this book given special attention.

Various battery technologies of interest for electric vehicles are described with a focus on the Li-ion battery technology. This is currently the most suitable technology combining high-energy density with high-power capability. Independent of the choice of technology, a battery consists of a number of cells connected in series and/or parallel, scaled according to vehicle type and purpose. This may sound like a very simple optimisation and design task, but to create a functional battery many more components are needed. Electric vehicle battery design requires skills in many different areas, e.g. materials science, electrochemistry, thermodynamics, dynamic modelling, control systems, electrical and mechanical engineering.

This book describes the underlying constraints needed to be understood for anyone aiming to design and/or use a battery for electric vehicles. This can be someone with her/his basic field of knowledge in, for example, chemistry, electronics, or physics. In addition, anyone familiar with battery chemistry and electrochemistry may want to be able to understand the possibilities of electric vehicles and vice versa; it may help those skilled in vehicle engineering to explore the field of electric vehicles and understand the possibilities and limitations of employing batteries.

Most of the fundamentals provided in this book can also be applied to the design of batteries for other applications e.g. large-scale stationary energy storages. Anyone with a general interest in battery materials or a specific interest in Li-ion batteries may also gain from the very practical and direct level of this book.

In order to provide a logical and distinct account of the process needed to accomplish optimised battery usage in electric vehicles, the book is divided into three main parts.

Part I explains the basics of electrochemical cells. The basic theories are described as well as the most common and suitable battery technologies for electric vehicles: lead-acid, nickel metal-hydride, and lithium batteries. Also capacitors and fuel cells are given brief attention. General electrochemistry applicable to battery cells is illustrated by schematic pictures.

In *Part II*, a more thorough review of the Li-ion battery technology is conducted. Materials for the different parts of the cell, as well as the complete cell, are treated. Battery functionality is described from its origin at the material level of the cell and how it is affected both by active and inactive materials. Material requirements, possibilities, and constraints are all included, and their influence on battery capacity, energy, and power. There is substantial discussion on how a cell can be optimised for either energy or power, and why the two cannot be combined in one single cell. Examples display how different active cell materials affect cell performance. The thermodynamic and kinetic properties of the active materials, factors determining cell performance, and the features of electrode design, e.g. particle size, electrode porosity and thickness, and cell format (cylindrical, prismatic, or pouch), all contribute towards final cell performance. Moreover, the manufacturing process of Li-ion cells is briefly described.

Finally, *Part III* focuses on the battery design for an electric vehicle. The different types of electric vehicles (all-electric, hybrids, and plug-in hybrids) are described. The battery requirements and the desirable operational conditions for the different types of electric vehicles are provided in general terms, including different energy and power requirements. How these requirements should be utilised depending on the battery design, including thermal management, is handled in the section on the battery management system. The state functions – state of charge, state of health, and state of function – the main input parameters to the battery management system, are described from a material perspective, i.e. the underlying mechanisms determining how to control a battery in an optimal way, both short-term and long-term. How to act to select the most optimal cell for a specific vehicle is described.

The last section of Part III takes an in-depth look at how cells age, mainly due to cell degradation, and how usage, usage history, and storage conditions affect the rate of degradation. Understanding of underlying degradation mechanisms provides a decision basis for battery design and development of control strategies. Finally, some methods of analysis, commonly used in order to understand how a battery degrades, are presented.

Most examples in the book are illustrated by the Li-ion battery technology, but the fundamentals are valid for all the battery types included. All figures throughout the book are made schematic rather than detailed to illustrate general trends and behaviours.

I

Electrochemistry and battery technologies

1 The electrochemical cell

The most fundamental unit of a battery is the electrochemical cell. All performance characteristics are dependent on the materials inside the cell, and all cells work according to some general principles independent of the materials employed. The purpose of this chapter is to bring together the fundamental aspects of an electrochemical cell as the basis for all further steps in the development of a battery intended for electric vehicles.

An electrochemical cell converts chemical energy to electric energy when discharged, and vice versa. In addition, the electrochemical cells can be said to be either *electrolytic* or *galvanic*. In an electrolytic cell, the electric energy is converted to chemical energy (charging of the battery) and in a galvanic cell chemical energy is converted to electric energy (discharging of the battery).

The basic design of an electrochemical cell consists of a positive and a negative *electrode* separated by an *electrolyte*, as shown in Figure 1.1. The chemical reactions taking place during charge and discharge processes are based on electrochemical *oxidation* and *reduction* reactions, known as the *redox reactions*, at the two electrodes. In these reactions, electrons are transferred via an external circuit from one electrode to another, and at the same time ions are transferred inside the cell, through the electrolyte, to maintain the charge balance. The species oxidised is called the *oxidant*, and the species reduced is called the *reductant*.

The oxidation reaction takes place at the negative electrode, the *anode*, and electrons are transferred, via the external circuit, to the positive electrode, the *cathode*, where the reduction reaction takes place by accepting the electrons. The negative electrode is thus an electron donor, and the positive electrode an electron acceptor. During charge and discharge of a battery, the nomenclature of the electrodes changes. Conventionally, the negative electrode is called anode and the positive electrode is called cathode, regardless of the cell being charged or discharged. Henceforth, however, only the terms positive and negative electrodes will be used, in order to simplify the discussion and avoid confusion.

Electrochemical cells are further classified depending on their ability to act both as electrolytic and galvanic cells. *Primary cells* are entirely of galvanic nature and can only convert chemical energy to electric energy. This type of cell is thus non-rechargeable and is therefore not of any interest for electric vehicle applications, and will not be discussed further. *Secondary cells* can operate both as galvanic and electrolytic cells, and are thus rechargeable. A fundamental understanding of the different processes occurring inside a secondary cell, which in turn depends on the materials used, is crucial for a proper understanding of the behaviour and performance of any electric vehicle.

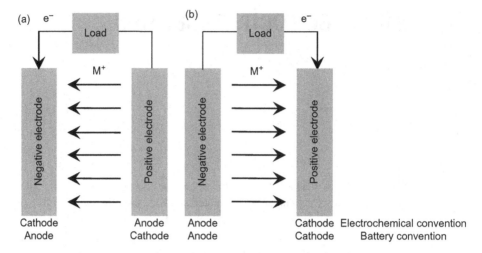

Figure 1.1 (a) An electrolytic and (b) a galvanic cell, respectively.

1.1   Definitions

A common language is important in any field of science, and also the electrochemistry and battery field has its own. To describe an electrochemical cell, some general rules are applied, and the notation is primarily based on the individual electrodes. The negative electrode is for shorthand notation always written to the left, and the positive to the right, and vertical lines mark the phase boundaries:

negative | electrolyte | positive

During discharge, electrons flow from left to right via the external circuit and ions flow in the same direction internally in the electrolyte, and naturally in the reverse directions during the charging process. This notation is used for the definition of the cell voltage, E_{cell}, which is always positive and defined as:

$$E_{cell} = E(right) - E(left) = E(+) - E(-) \tag{1.1}$$

After a first listing of the parts of the cell, the thermodynamics and the physical behaviour of a cell will be described in detail based on these very general notations.

1.2   Cell components

A cell consists of several components (Figure 1.2), all having a designated responsibility for the overall performance, possibilities, and limitations of the complete cell. There are both *active* and *non-active* components in a cell. The active components are those directly involved in the redox reactions of the cell and the non-active are those passive in the redox reactions, but can nevertheless be involved in other, side reactions taking place during charge and discharge of the cell. Generally speaking, the active

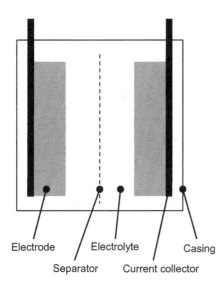

Electrode   Electrolyte   Casing

Separator   Current collector

Figure 1.2 The fundamental design of an electrochemical cell.

components are the electrodes, and the non-active components are everything else: the electrolyte, the separator, the current collectors, and the casing. The basic functions and requirements of each of these components are described separately below.

1.2.1   Electrodes

An electrode is an electrically and ionically conducting material most commonly being either metallic or of insertion type. In metallic electrodes, the electrochemical reactions take place only at the outermost surface layer, and these electrodes are therefore sometimes referred to as *blocking electrodes* due to the blocking of the surface by reaction products, limiting any further reactions. The most common electrodes for rechargeable cells are, however, insertion or *non-blocking electrodes*, where the redox reactions take place at the surface as well as in the bulk of the electrode. Blocking and non-blocking electrodes are illustrated in Figure 1.3. In addition, there are some electrochemical technologies (e.g. redox flow batteries, Section 2.5.5) that utilise electrodes where the redox couples are in solutions and the redox reactions take place at the interface between solutions. This type of electrodes will, however, not be discussed any further in this chapter.

1.2.2   Electrolytes

The electrolyte is a dynamic and essential part for the total performance of the electrochemical cell and has, despite its often very simple appearance, several crucial roles to play. Most electrolytes are a solution of one or several *salts* dissolved in one or several *solvents*. The electrolyte can be seen as the glue between the electrodes, but it also separates the electrodes to prohibit internal short circuit. To conduct ions and not electrons is, however the main task of the electrolyte. The ion conductivity should be

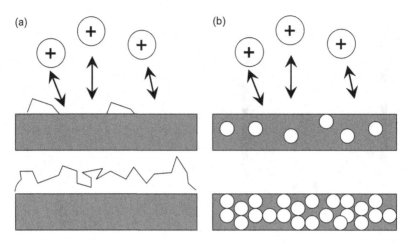

Figure 1.3 Blocking (a) and non-blocking electrodes (b) during redox reactions (top) and in a fully charged state (bottom).

fast in order not to limit the redox reactions in any operational condition. Generally, a liquid electrolyte has a high ion-conductivity, but low mechanical strength. On the other hand, polymer and solid electrolytes are more flexible in format and can be made in several shapes. The drawback is lower ion conductivity. Since the electrolyte should separate the electrodes, both the mechanical and electrical properties are important. The electrolyte must be compatible with the electrodes without losing its performance over time, and be stable within the full voltage range of the cell.

1.2.3   Separators

If a liquid electrolyte is employed in the cell, a separator material is often required to improve the mechanical properties of the electrolyte and thereby prevent any direct contact of the electrodes i.e. internal short circuit. The separator is most often a porous membrane soaked in electrolyte before the cell is assembled. The separator material must allow high ion conductivity by the electrolyte and maintain good electronic insulator properties. Therefore, in the selection of a separator material for a specific application and usage, several aspects must be considered, e.g. mechanical stability, mechanical strength, chemical stability, wetting of the electrolyte, and porosity. Often a fibre-structured material is used, and the shape and structure, as well as the porous structure, of the fibres can be tailored to obtain specific properties to meet specific requirements.

1.2.4   Current collectors

In order to secure the best possible cell charge and discharge processes, the electrons must be transferred from one electrode to the other in the external circuit as effectively as possible. Therefore, special components, the current collectors, are used at both electrodes. Materials with significant electrical conductivity, such as thin foils or grids of aluminium or copper, are used. In addition to their conductive properties, the current collectors must also be stable with respect to the electrochemical environment inside the cell, i.e. not take part in the cell redox reactions or be severely affected by any side reactions. During the operation of the cell, heat is unavoidably generated (Section 1.4.3) and the current collectors efficiently remove heat from the often more sensitive electrodes. Moreover, the current collectors also add mechanical strength, especially important in the case of non-metallic electrode materials, which often are based on nano-sized composites with consequently reduced mechanical performance.

1.2.5   Casing

The electrochemical cell is eventually included in a casing in order to create the final mechanical stability for the cell and limit any outside influence. The casing also protects the liquid electrolyte from simply evaporating. The casing can be made in different shapes and materials, e.g. plastic or metallic, hard or soft.

1.3   Cell and battery

There are basic design parameters to be considered in order to control the performance of a cell, and these are most often totally independent of the cell chemistry used. Primarily, the physical size and shape of the cell are important to find the optimal cell for any application. Secondarily, the configurations of the electrodes will affect the cell performance, e.g. the capacity. The basic design criterion for any cell, however, is how to arrange the two electrodes and the electrolyte. The exact design ultimately depends on the intended usage and performance demands, and a few basic types of cell designs are explained below, taking into account the application aimed at. These basic designs will be used further on to explain the operational constraints of a battery for electric vehicles.

1.3.1   Half cells

While not a cell design aimed at practical usage, quite often it is smart and of considerable advantage, especially in the R&D stages of cell development, to study the electrodes one at a time. This can be done by physically separating the full electrochemical operating redox cell into *half cells*. Within these, either the oxidation or reduction of the electrode of interest occurs and an electrode having a well-known and well-defined potential is used as the counter electrode. The most typical example is to study various positive or negative electrodes for Li-ion cells using a metallic lithium electrode as the counter electrode (while the final Li-ion cell of interest would employ e.g. a graphite negative electrode).

Depending on the electrodes used, various charge/discharge conditions become valid. In Figure 1.4, a comparison between a full cell and a half cell is illustrated. Regarding Li-ion half cells, the electrode of interest is always the positive electrode and the

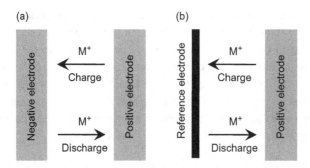

Figure 1.4 Full cell (a) and half cell (b).

metallic lithium electrode the reference electrode (the negative electrode in Figure 1.4). One of the main purposes is to try to understand how the lithium insertion into the electrode host structure occurs in detail and another is to accurately measure the lithium content as a function of the state of charge (Section 6.2.1) of the cell. Moreover, half cells are also commonly used in order to better understand the degradation processes occurring in aged cells. This is made via disassembling used cells and then reassembling each electrode into new half cells for post-mortem analysis (Section 7.3.6).

While half cells are very useful to decipher various phenomena separately and unambiguously, they do miss out on the important interactions involving more cell components in conjunction. For this, a full cell design is required.

1.3.2   Full cells: monopolar and bipolar

The most common way to construct an electrochemical cell is in a *full cell* design, i.e. a cell made of the two electrodes, fulfilling the performance requirements for the specific application. Depending on how the electrodes are arranged, either a *monopolar* or a *bipolar* cell design is accomplished.

The monopolar design is the more common. In such a design, each cell consists of one positive and one negative electrode and an electrolyte stacked in between, and the electrodes are connected outside the cell compartment. This stacking sequence can be continued in order to increase the cell capacity; in other words, there can be several positive and negative electrodes per cell. The main drawback with the monopolar cell design is the ohmic losses (Section 1.6.1) within the electrodes, leading to asymmetric current distribution since the flow of electrons takes place along the electrode surface area. The main advantage of this design is the generally high capacity, over a low power capability.

In contrast, in the bipolar cell design the positive and negative electrodes are assembled on opposite sides of an electron-conducting membrane, sandwiched in a serial configuration. The electrodes are thereby shared by two series-connected electrochemical cells; one side acts as the negative electrode in one cell and the other side as the positive electrode in the next cell. The electron-conducting membrane also acts as an

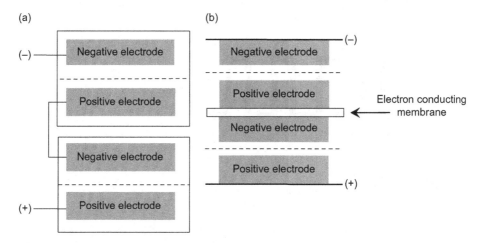

Figure 1.5 Comparison of the monopolar (a) and the bipolar (b) cell designs.

ion-diffusion barrier between the positive and the negative sides. In Figure 1.5, the monopolar and the bipolar cell designs are shown.

The design of the bipolar cell is made of stacked plates in series to obtain a pre-set given voltage and reduces the volume needed. The flow of electrons in the bipolar cell design is short, as it is directly through the comparatively thin membrane. The overall cell resistance is lower for the bipolar cell design compared to the monopolar due to a reduced number of connections between the cells, and the current distribution is more uniform over the electrode area. The main challenge for the bipolar cell design is to find suitable mechanically and chemically stable electron conductive membranes. The major failure modes of the bipolar design, which are not found in monopolar designs, are due to the cell sealing; current leakage between the cells will cause a decrease in cell performance, and pinholes in the electron-conducting membrane will cause electric short circuits.

1.3.3   Full cells: three-dimensional

The most commonly used cell designs, be it monopolar or bipolar, are always based on a two-dimensional concept. The extension of the cell design into three dimensions gives electrodes with larger accessible surface areas, and thereby shortens the distances for the ion transfer – together resulting in potentially higher volumetric energy densities and also power densities. The 3D cell design was primarily developed for small cells and micro-cells, i.e. cells of a few hundreds of mAh per cell for applications such as hearing aids. There are a number of ways to design 3D cells; one example is given in Figure 1.6. In a vehicle, vibrations are common, and if only this or other similar 3D cell designs can withstand these vibrations, it should be considered suitable for electric vehicles. Production techniques for large format cells are currently being improved, making it possible to develop a robust 3D cell design for cells suitable for vehicle application.

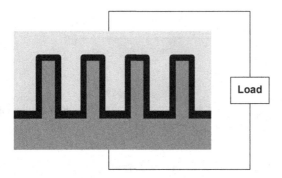

Figure 1.6 A representation of a 3D cell design.

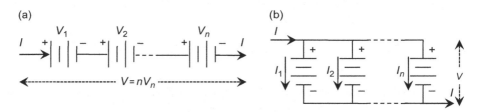

Figure 1.7 Series (a) and parallel (b) connected cells.

1.3.4   Battery

Electrochemical cells are limited in terms of voltage and capacity. Therefore, a battery intended for electric vehicles consists of many cells connected in series and/or parallel depending on the requirements in terms of voltage, power, and capacity.

Cells connected in series share the same current, I, while cells connected in parallel share the same voltage, V, as illustrated in Figure 1.7. If the cells are identical, the voltage of the series-connected cells is the sum of the voltages of all individual cells. In the case of identical cells connected in parallel, the current is divided equally among the individual cells, resulting in increased battery capacity.

1.4   Thermodynamics

To understand the behaviour, possibilities, and limitations of an electrochemical cell, attention must be given to the processes occurring inside the cell. Independent of the chemicals and the materials used, there are some fundamental physical relationships to be considered before assembling cells into a battery for an electric vehicle. The main properties of interest are voltage, capacity, energy, and power, which ultimately all originate in, and are limited by, the thermodynamics (theoretically) and kinetics (often in practice) of the chemical reactions.

The thermodynamic relationships are the basis for the general redox reactions and the temperature dependency of the electrochemical reactions in the cell. For example, the thermodynamics of the materials selected for the cell determine the maximum voltage level possible and the (ideal) behaviour during charge and discharge. How fast the chemical energy can be transformed to electrical energy depends on the mass transport and diffusions within the different parts of the cell, and is also strongly dependent on the materials used.

Based on thermodynamics, it is the potential difference between the electrodes that determines the theoretical *cell voltage*, E_{cell}, also referred to as the *electrochemical force* (emf). This is primarily expressed as Gibbs free energy of the charge of the cell, ΔG_{cell}. At *open circuit*, i.e. no current flowing, there are no reactions taking place and there is an electrostatic force arising from the potential difference between the electrodes, which must be balanced by a chemical force. When the reaction occurs, there is a decrease in the free energy of the cell according to:

$$\Delta G_{cell} = -nFE_{cell} \tag{1.2}$$

where F is the Faraday constant (=96485 As/mol), and n is the number of electrons involved in the reactions. E_{cell} is related to the sum of the standard potential of each electrode according to:

$$E_{cell} = E_{positive} - E_{negative} \tag{1.3}$$

This can be illustrated by an example using a cell of the negative electrode A and the positive electrode B, where the electrolyte is conducting A^{n+} ions between the electrodes, and the new compound AB is formed at the B electrode. The following reactions take place:

$$\text{Negative electrode: A} \rightarrow A^{n+} + ne$$

$$\underline{\text{Positive electrode: } A^{n+} + ne^- + B \rightarrow AB}$$

$$\text{Net cell reaction: A + B} \rightarrow AB$$

Considering the change in Gibbs free energy for any chemical reaction, the total ΔG_{cell} for the electrochemical cell is the difference between the sum of free energy of the products and the sum of free energy of the reactants according to:

$$\Delta G_{cell} = \sum \Delta G_{prod} - \sum \Delta G_{react} \tag{1.4}$$

If $\Delta G_{cell} > 0$, energy is required to fulfil the electrochemical reaction, i.e. the electrochemical force $E_{cell} < 0$, and corresponds to an electrochemical cell during charge. On the other hand, during discharge the reaction takes place spontaneously by converting chemical energy to electrical energy, i.e. $\Delta G_{cell} < 0$ and $E_{cell} > 0$.

1.4.1   Chemical and electrochemical potentials

The different redox reactions can also be expressed in terms of chemical and electrochemical potentials in order to further understand the correlation between cell behaviour and the physical constraints of the cell. Consider the same electrochemical net cell reaction as above:

$$A + B \rightarrow AB$$

For this redox reaction, the change in Gibbs free energy can also be presented in terms of chemical potential:

$$\Delta G_{cell} = \sum \mu_{prod} - \sum \mu_{react} = \mu_{AB} - (\mu_A + \mu_B) \qquad (1.5)$$

where μ_i is the chemical potential of species i. The chemical potential defines the change in Gibbs free energy when an infinitesimal number of moles of species i is added with all other components remaining constant. Moreover, the chemical potential is dependent on the activity a of the species i according to:

$$\mu_i = \mu_i^0 + RT \ln a_i \qquad (1.6)$$

where μ_i^0 is the standard chemical potential for species i, R is the gas constant, and T is the temperature in Kelvin.

Furthermore, if the negative electrode A is made of metal and brought into contact with A^{n+} ions in the electrolyte, the following reaction will occur until equilibrium is reached:

$$A^0 \text{ (metal)} \leftrightarrow A^{n+} \text{ (electrolyte)} + ne^- \text{ (metal)}$$

This leads to a charge separation between the electrolyte and the solid metal electrode and occurs spontaneously at equilibrium. The charge accumulation on each side of the interface between the electrode and the electrolyte can be described in terms of an electrostatic potential in the electrode, $\phi_{electrode}$, and in the electrolyte, $\phi_{electrolyte}$, together creating an electrical potential difference, $\Delta\phi$, referred to as the *Galvani potential difference*:

$$\Delta\phi = \phi_{electrode} - \phi_{electrolyte} \qquad (1.7)$$

The chemical potential is related to the chemical equilibrium between the different species i, without any charge separation at the interface. However, the electrostatic potential influences the chemical potential of an electrochemical cell; the *electrochemical equilibrium* has to consider the electrostatic potential ϕ in each phase. The electrochemical potential, $\bar{\mu}_i$, can therefore be defined as the combination of the chemical potential and the Galvani potential according to:

$$\bar{\mu}_i = \mu_i^0 + RT \ln a_i + nF\phi_i \qquad (1.8)$$

1.4.1.1   Nernst equation

The electrochemical potential is more or less directly related to the cell potential. In the case of an electrochemical cell with two electrodes having different potentials and different activities of species i, a_i, and both being ruled by equation (1.8), the electrochemical potential can be expressed as:

$$\Delta\bar{\mu}_i = \bar{\mu}_{i,+} - \bar{\mu}_{i,-} = RT(\ln a_{i,+} - \ln a_{i,-}) + nF\phi_i = RT\left(\frac{a_{i,+}}{a_{i,-}}\right) + nF\phi_i \qquad (1.9)$$

The activity of a pure solid phase is by definition 1, and the activity of the electrons in the electrode will not be affected by establishing the electrochemical phase equilibrium (due to the large electron concentration in the electrode). Combining the chemical and electrochemical potentials (equation (1.9)) with Gibbs free energy (equation (1.5)) results in the *Nernst equation*, where the voltage, E_{cell}, of a cell is given by:

$$E_{cell} = E^0 - \frac{RT}{nF}\left(\frac{a_+}{a_-}\right) \qquad (1.10)$$

The E_{cell} of equation (1.10) is the same as in equation (1.1). To practically measure E^0 of a specific electrode is considered impossible. Therefore, in order to establish a scale of half cell standard potential, a reference potential has been agreed and set as zero. By convention, 'zero' is defined as the standard potential of the $H_2/H^+_{(aq)}$ reaction – the *standard hydrogen electrode* (SHE). In Li-ion battery nomenclature, however, the cell potential is most often referred directly (and of course much more conveniently) to the Li^0/Li^+ potential, which in turn is –3.045 V vs. $H_2/H^+_{(aq)}$.

As can be seen from the Nernst equation, the cell potential is dependent on the temperature and the number of electrons involved in the redox reactions. Not only the cell potential is related to the temperature, the overall cell performance is strongly temperature dependent.

1.4.2 Cell voltage

The cell voltage is the potential difference between the two electrodes and is measured in volts (V). It is limited by the theoretical cell voltage, but is not necessarily the same. Electrochemical cells can have different cell voltages. For example, a Pb-acid cell has a cell voltage of 2 V, a NiMH cell has a voltage of 1.2 V, and a Li-ion cell a voltage of about 3–4 V – the latter depending on the exact chemistry selected. Knowing the cell voltage, and how the voltage changes during operation, is one of the key tasks in the battery design process.

The theoretical cell voltage is determined by the Nernst equation (1.10). Under non-load conditions, the cell has an *open circuit voltage* (OCV), which usually is close to the theoretical voltage. The typical rated cell voltage is the *nominal voltage*. The cell voltage changes as a function of usage, and the curve obtained is referred to as the *discharge profile* or the *voltage profile*. Typical discharge profiles are illustrated in Figure 1.8.

The shape of the discharge profile depends on both the cell technology and the materials used. The shape is ideally determined by the *Gibbs phase rule* (equation (1.11)) – in short, thermodynamics turned into electrochemical practice. For a defined electrochemical cell, there are a number of thermodynamic parameters or degrees of freedom, F, defining the system: temperature, pressure, and chemical composition/potential. In the normal case of constant temperature and pressure

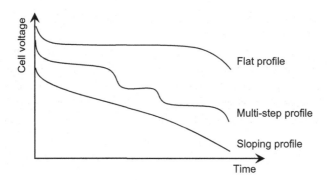

Figure 1.8 Typical discharge profiles for an electrochemical cell.

conditions, the only parameter able to vary is the chemical composition. Hence, F varies by the number of components, C, and the number of phases present in the active electrode material, P:

$$F = C - P + 2 \tag{1.11}$$

The degrees of freedom provide the number of parameters that can be varied during charge/discharge of the cell: temperature, pressure, atomic ratios, and voltage. Normally, temperature, pressure, and atomic ratios are all constant, and therefore only the voltage can change. There are two possible shapes of voltage profiles at equilibrium: flat or sloping. In the first case, the voltage is independent of the composition, and, in the latter, the potential is dependent on the composition, as illustrated in Figure 1.9. In this figure, the nature of a single electrode is shown, based on half cells. Therefore, only the electrode potential is given, not the cell voltage.

To illustrate how the Gibbs phase rule is valid in practice, some examples including commonly used electrode materials for Li-ion cells are presented here.

(1)   Lithium: a pure metallic electrode
    $Li^+ + e^- \leftrightarrow Li^0$
    $C = 1$ (only one component is present: Li)
    $P = 1$ (only one phase is active: Li)
    $\Rightarrow F = 2$
    Since temperature and pressure are fixed, no other parameters can be varied to define the system and the resulting profile will be a flat plateau.

(2) $LiFePO_4$: a two-phase ordered reaction of insertion materials
    $LiFePO_4 \leftrightarrow FePO_4 + Li^+ + e^-$
    $C = 4$ (Li, Fe, P, O)
    $P = 2$ (two phases are active in the redox reaction: $LiFePO_4$, $FePO_4$)
    $\Rightarrow F = 4$
    Since temperature, pressure, Fe:O and Fe:P ratios are fixed, no other parameters can be varied to define the system and the resulting profile will be a flat plateau.

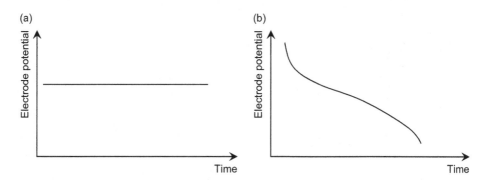

Figure 1.9 Shape of the voltage profile: (a) independent of composition and (b) dependent on composition.

(3) Graphite: a single-phase random diffusion reaction of insertion materials

$C + xLi^+ + xe^- \leftrightarrow Li_xC$

$C = 2$ (Li, C)

$P = 1$ (one phase is active, continuous insertion of Li in C)

$\Rightarrow F = 3$

Since temperature and pressure are fixed, one parameter can be varied to define the system and the resulting profile will be a sloping profile.

(4) LiCoO$_2$: a single-phase random diffusion reaction of insertion materials

$LiCoO_2 \leftrightarrow Li_{1-x}CoO_2 + xLi^+ + xe^-$

$C = 3$ (Li, Co, O)

$P = 1$ (one phase is active, continuous insertion of Li in $Li_{1-x}CoO_2$)

$\Rightarrow F = 4$

Since temperature, pressure, and the Co:O ratio are fixed, one parameter can be varied to define the system and the resulting profile will be a sloping profile.

For a general electrochemically active electrode material, reacting over a wide range of chemical compositions, the shape of the profile follows the phase diagram, as illustrated in Figure 1.10. Depending on the phase diagram, the voltage profile can include several different plateaus and sloping profiles in a multi-step character.

In this tentative electrochemical example, only one electrode is studied, i.e. a half cell is used. At its charged state and at temperature T, the electrode only contains compound A. As the discharge proceeds the α phase starts to form, and the cell potential decreases due to the solid solution character of the α phase (F>0). As Point 1 is reached, the potential is independent of the composition since the α and β phases coexist (F=0) until the discharge procedure has reached Point 2, where only the β phase exists, and the voltage starts to decline again. At Point 3, the voltage reaches the next plateau, the coexistence of the β and γ phases, and at Point 4 the solid solution of the γ phase correlates to a voltage decline until the pure compound B has been reached.

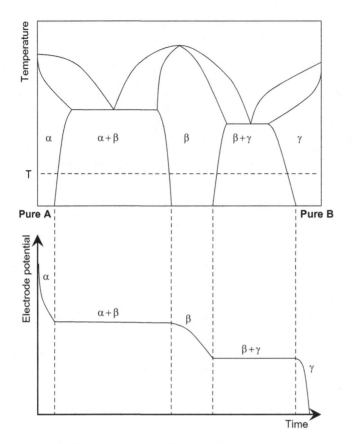

Figure 1.10 The shape of the equilibrium voltage profile in relation to the corresponding phase diagram.

1.4.3   Temperature

At higher temperatures than T, the same cell as above would result in a different voltage profile and at some point the electrode will be liquid. Depending on the nature of the phase diagram, and the temperature dependency, the voltage profile can vary with merely small differences in temperature and the resulting cell can be impossible to use. Thus, it is understandable that fundamentally the performance of an electrochemical cell is highly dependent on the operation temperature. The temperature also affects many other fundamental parameters, such as chemical activity and internal resistance. Utilisation of a cell – i.e. during vehicle operation, the performance of the cell, and thereby the vehicle – is strongly dependent on the temperature, the internal cell temperature, as well as the ambient. The internal temperature will increase due to the nature of the electrochemical redox reactions involved, and can also be increased by the ambient temperature. The rate of chemical reactions, including the side reactions, is increased by temperature. If the temperature increases too fast or becomes too high, the cell may exhibit conditions of an abuse situation, as a result of decomposition of materials and dominant side reactions. Therefore, all cells should be kept within specific temperature limits, often of a narrow range between 20–40 °C, and, in addition, a cooling/heating system is most often required in electric vehicles to optimise both usage and safety (Section 5.2.3.1).

Apart from the discharge profile above, the cell voltage is dependent on the temperature according to the Nernst equation (1.10), and the change in Gibbs free energy is proportional to the entropy change during the reaction:

$$\Delta G = -nFE_{cell} = \Delta H - T\Delta S = \Delta H - nFT\left(\frac{dE_{cell}}{dT}\right) \tag{1.12}$$

The term (dE_{cell}/dT), the entropy heat, is positive if heat is generated during charge, and hence the reverse reaction will consume heat during discharge. As the electrochemical reactions, however, are not fully reversible, an irreversible heat, q, will be generated during operation of an electrochemical cell according to:

$$q = i(E_{OCV} - E_{cell}) \tag{1.13}$$

where i is the applied current. This irreversible heat generation originates from the non-equilibrium operation of the cell and is released inside the cell mainly at the redox reaction sites, i.e. at the surface of the electrodes, and will cause losses. These losses during cell charging/discharging will generate heat – *Joule heating* – and can be dominant during cell usage. The total heat generated during discharge of an electrochemical cell is the sum of heat generated by the entropy heat and the irreversible heat released during cell operation. In relation to the irreversible heat generated, the contribution of the entropic change is relatively small compared to the overall increase in cell temperature. At low discharge rates, the heat release can be negligible, but at higher rates the heat released can cause severe cell damage. This must be taken into account when designing the cell, the battery, and the thermal management system of the battery; otherwise, the temperature increase may result in abuse situations, such as thermal runaway (Section 6.1.2.1).

Moreover, the temperature influences the capacity of the cell. Discharging at lower temperatures will result in a reduction of the capacity since the *cut-off voltage* will be reached faster due to the higher internal resistance. At elevated temperatures, the internal resistance decreases and results in increased capacity output.

1.5 Electrode and electrolyte processes

So far, we have almost exclusively considered the overall cell characteristics at the macroscopic level without any detailed attention being given to the microscopic level and the fundamentally limiting processes. In any electrochemical cell, however, the interactions between the materials, especially at all interfaces, are of uttermost importance for the final cell performance characteristics, e.g. power capability and durability. It is often at the interfaces between the electrodes and the electrolyte that most limiting

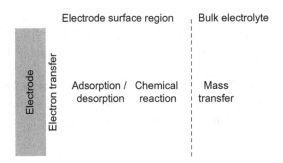

Figure 1.11 The electrode and electrolyte regions and the reaction paths at the electrode surface.

processes occur and in addition many side reactions have their origin. In the cell, the chemical reaction rates and how fast the different physical processes occur in the electrodes and the electrolytes also affect overall cell performance. Therefore, mass transfer, diffusion, and electron transfer are processes of major importance and will be thoroughly described next.

As a starting point, the electroactive species, i.e. the ions, are transported to the electrode by migration and/or diffusion and are adsorbed at the surface prior to electron transfer. Figure 1.11 illustrates the reaction paths near the electrode surface. It is in this region, near the electrode surface, that the electrochemical reactions take place through a charge-transfer process, characterised both by chemical and electrical changes. The bulk electrolyte, as well as electrode surface regions, should therefore be as small as possible if rapid reactions, and thereby high-power outputs, are preferred.

1.5.1   Electrode kinetics

The speed of the electrode reactions, the electrode kinetics, determines how quickly the chemical energy can be converted to electrical energy, i.e. the power capability of the cell. Electrode reactions are never infinitely fast and they all require a driving force to sustain a given rate. The rate at which the chemical energy is converted into electrical energy is expressed as *current*. At equilibrium, the anodic and the cathodic currents are equal in magnitude and correspond to the *exchange current*, i_0. In general, anodic currents are defined as positive currents and cathodic as negative. During cell operation, non-equilibrium states are reached either in a positive or a negative direction, and are accompanied by a deviation of the electrode potential from the equilibrium potential. This deviation, known as the *overpotential, η*, is described in Section 1.6.1. The relationship between the applied/received current, i, and the overpotential, η, is given by the *Butler–Volmer equation*:

$$i = i_0 \left(e^{\frac{\alpha n F}{RT}\eta} - e^{-\frac{(1-\alpha)nF}{RT}\eta} \right) \tag{1.14}$$

where n is the number of electrons involved in the electrochemical reaction and α is the *transfer coefficient*: the fraction of the electrode potential involved in the reaction. For small overpotentials, the Butler–Volmer equation is approximately linear according to:

$$i = \frac{i_0 F}{RT} \eta \qquad (1.15)$$

This equation can be further approximated as Ohm's law:

$$\frac{\eta}{i} = \frac{RT}{i_0 F} \equiv R_{ct} \qquad (1.16)$$

where R_{ct} is the *charge-transfer resistance*.

1.5.2 Electrode–electrolyte interfaces

For rechargeable electrochemical cells, the redox reactions are reversible, but the reaction rates for the oxidation and the reduction reactions may differ. These can depend on the concentrations of the reactant species at the electrode surfaces, which, in turn, often differ from the electrolyte bulk concentrations. For any electrode in contact with an electrolyte, an excess of charge is accumulated at the electrode surface and is counterbalanced by an accumulation of ionic species in the electrolyte close to the electrode surface. This interfacial region is called the *electrochemical double-layer*. The driving forces for its formation are primarily Coulombic, but also chemical interactions with the electrode surface, which determine the distribution of ions within the double-layer (Figure 1.12).

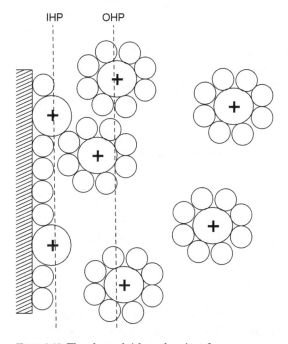

Figure 1.12 The electrode/electrolyte interface.

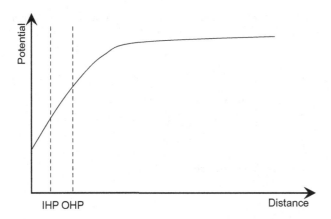

Figure 1.13 The potential distribution across the Helmholtz planes.

As illustrated in Figure 1.12, the ions in the electrolyte form a layered structure where they are often surrounded by solvent molecules. A number of models are applied to describe this layer, and one of the more widely used, due to its simplicity, is based on the theories developed by Helmholtz, to illustrate the double-layer characteristics. The layer closest to the electrode surface is formed by ions adsorbed and strongly interacting with the electrode surface – the *inner Helmholtz plane* (IHP). Ions found a further distance from the electrode surface are only weakly or electrostatically interacting with the electrode surface and tend to keep their solvent shell from the bulk electrolyte intact and are consequently the *outer Helmholtz plane* (OHP). Further out in the bulk of the electrolyte, more diffuse layers can be found (highly dependent on the local salt concentration). The potential distribution across the Helmholtz planes and further out in the bulk of the electrolyte is schematically shown in Figure 1.13.

The electrochemical double-layer is by nature a capacitor, where the electrode surface and the OHP act as the capacitor planes. Due to the narrow distance between the planes, the *double-layer capacitance*, C_{dl}, can be quite large (as the capacitance is always inversely proportional to the distance between the planes). The double-layer capacitance acts in parallel with the Butler–Volmer equation (1.14) and the relationship between the current at the double-layer, i_{dl}, and the polarisation, η, is:

$$i_{dl} = C_{dl}\frac{\partial \eta}{\partial t} \qquad (1.17)$$

This influences the dynamics of the electrochemical cell; the power rates of the cell will be limited due to increased polarisations and the losses will increase.

1.5.3　Mass transport

Mass transport by diffusion processes is the dominant process in the electrolyte for most electrochemical cell types suitable for vehicle applications. Improvement of the diffusion processes is thus essential in order to improve battery performance. More

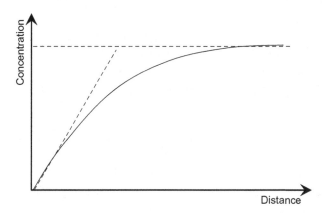

Figure 1.14 The concentration gradient from the electrode surface to the bulk electrolyte.

specifically, the mass transport to and from the electrode surface is essential for cell operation and is mainly controlled by diffusion of species in a concentration gradient. The diffusion follows *Fick's first law*, where the flux of ions, q, across the diffusion gradient distance, x is proportional to the concentration gradient according to:

$$q = D\frac{\partial C}{\partial x} \tag{1.18}$$

where D is the diffusion coefficient and C is the electrolyte concentration.

The flux of ions will limit the current and minimise the polarisation at the double-layer. The thickness of the concentration gradient layer depends on the electrolyte concentration. Figure 1.14 illustrates the concentration gradient near the electrode surface. At this level, the effects of the IHP and OHP, discussed above, are negligible.

This gradient results in a *concentration polarisation, η_c*, or overpotential, which is the same concentration polarisation as will be discussed in Section 1.6.1. According to the Nernst equation, this polarisation across the diffusion layer can be expressed as:

$$\eta_C = \frac{RT}{nF}\ln\frac{C_b}{C_s} \tag{1.19}$$

C_b and C_s refer to the concentration of ionic species in the bulk of the electrolyte and at the surface, respectively. Consequently, the polarisation losses will increase with the salt concentration.

1.5.4   Ion transport

In the bulk of the electrolyte, processes other than those at the electrode/electrolyte interfaces are dominant and affect cell performance. The bulk of the electrolyte is more dynamic, compared to the more or less steady state conditions near the electrode surface. In order to convert chemical energy into electric energy, ions have to be transported all across the electrolyte from one electrode to the other requiring retention of a good electrode/electrolyte contact during charge and discharge. This mobility, or *ionic conductivity* is another aspect of the cell current. The corresponding changes in the electrolyte composition depend on the position in the cell and the time. In the bulk, the changes in concentration are caused by diffusion (and migration), and as a first approximation *Fick's second law* can be applied:

$$\frac{\partial C}{\partial t} = D\frac{\partial^2 C}{\partial x^2} \tag{1.20}$$

The total ion transport is dependent on the processes at both the electrode/electrolyte interfaces, where the ions enter or leave the electrolyte by being (de)solvated by the solvent, and the solvated ions diffusing in the bulk of the electrolyte under the influence of the cell potential (*migration*). The resulting ion transport properties are mainly described in terms of *ionic conductivity, ion mobility*, and *transference number*; properties directly related to the current density and the applied electric field.

The ionic conductivity, σ, often measured as a current, depends on the ion mobility and the number of charged species involved in the electrochemical reactions according to:

$$\sigma = \sum_i n_i \mu_i z_i e \tag{1.21}$$

where n_i is the number of charge carriers, μ_i is the ion mobility, z_i is the valence of the species, and e is the unit charge.

As a first approximation, the ion mobility follows the *Stokes–Einstein equation*:

$$\mu_i = \frac{1}{6\pi\eta r_i} \tag{1.22}$$

where r_i is the solvation radius and η is the viscosity of the electrolyte. The unit of the ion mobility is $\Omega^{-1}cm^2$ or $V^{-1}s^{-1}cm^2$. The Stokes–Einstein equation can also be written as:

$$D_i = \frac{RT}{N_A}\frac{1}{6\pi\eta r_i} \tag{1.23}$$

N_A is Avogadro's number, R the gas constant, and T the temperature. This thus connects equations (1.20)–(1.22) in a single equation.

The transference number is defined as the fraction of the total current carried by a specific ion of interest:

$$t_i = \frac{i_+}{\sum_i i_i} \tag{1.24}$$

where $t_+ + t_- = 1$. In the case of a Li-ion or a NiMH cell, the t_{Li}^+ and t_{OH}^-, respectively, are of interest in order to obtain rapid ion transfer and to maximise fast power response.

Increased ionic conductivity improves the overall performance of the electrochemical cell by, for example, allowing faster reactions. The bulk ionic conductivity can be

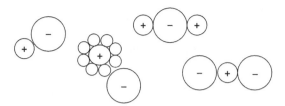

Figure 1.15 Ion pairs and higher aggregates.

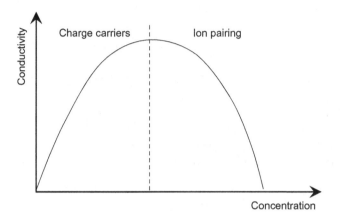

Figure 1.16 The ion conductivity as a function of salt concentration.

increased by decreasing the viscosity equation (1.22) and as the viscosity is inversely proportional to the temperature, raising the temperature is a direct way to increase the ion conductivity. To increase the transference number of ions involved in the electrochemical reactions, however, the transference number of the counter-ions must be decreased, e.g. by choosing larger counter-ions to lower their mobility – in practice by selecting another salt for the electrolyte.

Only charged species migrate in the electric field created by the cell potential and contribute to the ionic conductivity. Therefore, the creation of associated and neutral ion pairs, *ion pairing*, is a phenomenon unwanted in electrochemical cells. As the number of ion pairs increases, the number of charge carriers, n, will decrease, and hence a decrease in the ion conductivity will be observed (equation (1.21)). In addition, higher aggregates may be formed, and these species decrease the ionic conductivity by an impeded ion mobility, μ_i. In non-aqueous electrolytes, the numbers of ion pairs (and aggregates) grow with temperature due to the balance between enthalpy and entropy contributions to the salt solvation. Figure 1.15 illustrates ion pairs and higher aggregates.

The ion conductivity depends on the salt concentration, the temperature, the viscosity, and the solvent used. Figure 1.16 shows an example of how the ion conductivity depends on the salt concentration. At low concentrations, the number of charge carriers is the limiting factor and at high concentrations ion pairing is a strongly limiting process, which decreases both n and μ (equations (1.21) and (1.22)). In general, the

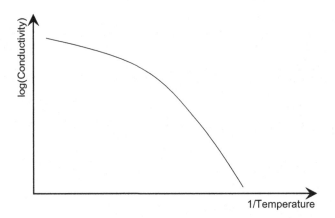

Figure 1.17 Arrhenius plot of ion conductivity as a function of temperature.

maximum ion conductivity is therefore found at intermediate salt concentrations, often *ca.* 1 M, balancing these two limitations.

For a liquid electrolyte, the influence of temperature on the ion conductivity is the non-linear behaviour that follows the Arrhenius equation:

$$\sigma = \frac{A}{T}\exp\left(-\frac{E}{kT}\right) \tag{1.25}$$

where k is the Boltzmann constant, A is a constant, and E is the activation energy, and is illustrated in Figure 1.17.

1.5.5   Mass transport in solid states

The mass transport within the electrodes is to a large extent responsible for the ability of the cell to deliver the requested energy within a reasonable time frame. Mainly two classes of electrodes are of interest here: porous or planar (i.e. non-blocking or blocking). Most rechargeable electrochemical cells applicable for electric vehicles comprise porous electrodes, made of more or less crystalline materials, whereas planar electrodes are made of pure metals, e.g. Li(s) (Section 3.1.2).

The mass transport is affected by a number of parameters: particle size, morphology, pore structure, thickness, and mass loading of the electrodes, resulting in different cell performance characteristics, e.g. power or energy-optimised cells. A porous electrode has a larger surface area available for the electrochemical reactions, and provides several pathways for current flow. Moreover, the interface between the electrode and the electrolyte is larger for the porous electrode, enabling better electronic and ionic transfer between the two. Figure 1.18 shows a schematic representation of a porous electrode. Besides the electrochemically active electrode material, a real composite electrode also comprises electron-conducting particles, e.g. fine carbon particles, a chemical inert binder, and electrolyte, all attached to a current collector. In order to achieve the required electrochemical performance, the pores must be filled with electrolyte to provide the charge transfer.

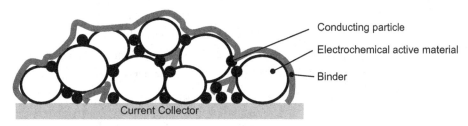

Figure 1.18 Schematic picture of a porous electrode.

(a)   (b)   (c)

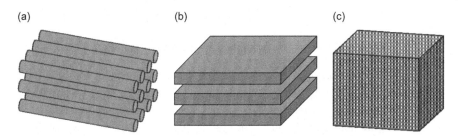

Figure 1.19 Schematic illustration of diffusion pathways in electrode materials; (a) 1D, (b) 2D and (c) 3D.

1.5.5.1   Insertion and conversion materials

The active material used in porous electrodes foremost belongs to two basic families of materials, *insertion* and *conversion* materials; hence the naming of the different electrode types.

Insertion materials utilise reactions involving guest species entering unoccupied sites of a host without any major changes in the structure upon charging or discharging. Sometimes this mechanism is referred to as *intercalation*, but in the following only insertion will be used. Two main characteristics distinguish insertion materials from other solids: (i) the guest species is moving between sites in the host structure, and (ii) the concentration of the guest species can vary. This type of reaction provides adequate cyclability, i.e. charge and discharge of the cell, but usually only a relatively low concentration of guest species can be incorporated in the host, resulting in a low capacity.

Indeed, for all insertion materials the capacity is limited by the number of available interstitial sites in the host structure that can be occupied by the guest species. The sites available depend on the structure of the host, and different diffusion pathways are available between the sites, connected as one-, two- or three-dimensional structures, as illustrated in Figure 1.19. The different diffusion pathways give rise to different electrode properties and thereby to different cell performance constraints for 1D, 2D, and 3D structures. If the pathway were suddenly blocked in a 1D structure, the charging/discharging of the cell would be limited since the inserted ions would not be able to enter/leave the structure. On the other hand, for 2D and 3D structures, the insertion of the guest species may result in volume expansion in one, two, or three dimensions, resulting in mechanical stress. It is thus crucial how the porous electrode is designed to handle these volume changes.

The insertion of the guest species into the host structure can be achieved in several ways and involves one of the following reaction schemes:

- single-phase insertion/solid solution reaction with a continuous change of phase composition by a gradual insertion of the guest species:

$$xA + BC \leftrightarrow A_xBC \text{ where } 0 \leq x \leq a$$

$$\text{example: } xLi + CoO_2 \leftrightarrow Li_xCoO_2$$

- two-phase insertion reaction involving two phases having different concentrations of guest species:

$$xA + yB \leftrightarrow A_xB_y$$

$$\text{example: } Li + FePO_4 \leftrightarrow LiFePO_4$$

Conversion materials, on the other hand, have reactions that comprise the formation of new phases during charge and discharge. This causes large mechanical stress in the electrode, and the reactions are often of a two-phase character. These electrodes are therefore not easily designed for cyclability even if the capacity may be high. Mainly two ways exist to create these new phases without involving any insertion of guest species, the new phases being formed by the disappearance of other(s):

- formation reaction, where a new phase is formed from the basic elements:

$$A + B \leftrightarrow AB$$

$$\text{example: } Sb + Li \leftrightarrow LiSb$$

- conversion reaction, where one compound is displaced from the host structure and replaced by another:

$$A + BC \leftrightarrow AC + B$$

$$\text{example: } SnSb + 3Li \leftrightarrow Li_3Sb + Sn$$

Single-phase reactions, that only occur for insertion electrodes, are of a random order character, i.e. the inserted species are entering sites without any structural ordering. The two-phase reactions occurring for both insertion and conversion electrodes, however, can be based on either random or ordered diffusion of the formation of the new phase. This is illustrated in Figure 1.20.

Even if the atomic structure of the host is almost unaffected by insertion, the electronic structure will change and this ultimately determines the theoretical potential. The bonding and anti-bonding electronic levels are arranged in bands, and this set-up is schematically shown in Figure 1.21. The electrons from the guest species are added to the conducting bands as the insertion process proceeds and the Fermi level (E_F) of the host changes to a higher level. The insertion process only proceeds until the conducting band has been filled, and the oxidation state of the host structure has reached its limit, and capacity is related to the number of inserted species. Often the

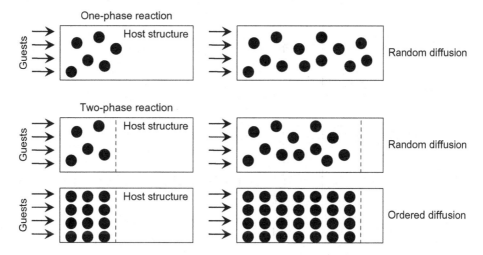

Figure 1.20 Reaction schemes for the evolution of insertion and conversion reactions, of single or two-phase character, in electrode materials.

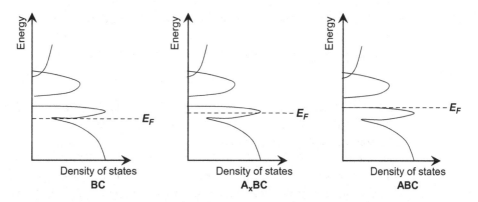

Figure 1.21 The changes in the electronic band structure of a solid host during an insertion process.

host structure involves transition metals changing their oxidation state during the insertion/extraction reactions, and the reaction is terminated as the highest/lowest oxidation state is reached.

1.5.6   Electrolyte stability

Chemical stability of an electrochemical cell requires that the electrolyte is neither reduced at the negative electrode nor oxidised at the positive electrode. Thermodynamic stability of the electrolyte is only achieved if the Fermi level of the negative (E_{Fneg}) and the positive (E_{Fpos}) electrodes are matched with the LUMO (lowest unoccupied molecular orbital) and HOMO (highest occupied molecular orbital) levels of the electrolyte according to the schematic illustration in Figure 1.22.

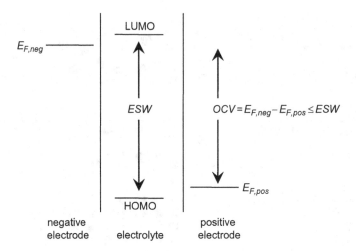

Figure 1.22 Placement of electrode (mixed valence redox couple) energies relative to the electrolyte.

The electrolyte must remain within this *electrochemical stability window* (ESW) in order to maintain thermodynamic stability. An E_{Fneg} above the electrolyte LUMO will reduce the electrolyte, and an E_{Fpos} below the electrolyte HOMO will oxidise the electrolyte. Therefore, thermodynamic stability restricts the OCV to:

$$OCV = E_{Fneg} - E_{Fpos} \leq ESW \qquad (1.26)$$

In most electrochemical cells, however, the electrolyte is not stable within the full voltage range of the cell, and side reactions will take place due to reduction and/or oxidation reactions of the electrolyte. The ESW is, in fact, determined by its ability to make use of these side reactions in order to create a passivation layer on the electrode and thus blocking any further electrolyte reduction or oxidation. Meta-stability or kinetic stability is created for a limited extra voltage range and the ESW will this way be extended. A passivation layer for the Li-ion cell is often referred to as a solid-electrolyte interphase (SEI; Section 3.1.1), a phase allowing both ion transfer and electron transfer during closed-circuit conditions.

1.6   Practical cell measures

Up to now, the basics and the underlying theories of an electrochemical cell have been described in some detail. Understanding the many possibilities, essential limitations, and basic performance characteristics of the cell is the foundation, but it is now time to focus on understanding practical cell behaviour during load/usage.

The properties of interest when designing a battery for an electric vehicle are mainly the cell voltage and how it changes during operation, the capacity of the cell, and how the energy and power can be utilised in the most optimal way. From an electric vehicle perspective, the capacity of the electrochemical cell corresponds to the size of the fuel tank in a conventional vehicle running on gasoline or diesel. The energy content refers to the distance an all-electric vehicle can drive, and the power determines the acceleration or energy recuperation abilities. Hereinafter, these parameters and how the performance characteristics are affected by the load constraints will be discussed.

1.6.1   Cell voltage under load

The picture given of the OCV (e.g. Figure 1.8) is by definition only valid for cells at rest and not for those in operating condition. When the cell is employed in an application, the voltage will thus differ from the OCV. During discharge, the cell transforms its chemical energy to electrical energy. At the same time, the voltage drops below OCV due to losses in the cell. These losses are caused by various types of polarisation, or overpotential, occurring when a load current i passes through the cell. The polarisation types are generally described to be of activation, concentration, or ohmic character.

Activation polarisation originates from the electrochemical reactions at the electrode surface, and can be seen as losses due to charge transfer, η_{ct}. *Concentration polarisation, η_c,* arises from concentration differences of charged species between the electrode surface and the concentration in the bulk of the electrolyte. This polarisation is due to limitations in the mass transfer from/to the electrode/electrolyte interface, and hence by the ionic conductivity and transport properties of the electrolyte.

The polarisation can be calculated based on electrochemical and mass-transfer data. It is difficult to separate the activation and concentration polarisations by measurements, and a sum of the two is normally obtained. Most electrode materials in modern batteries are made of composites having porous structures, resulting in complex mathematical models in order to estimate the polarisation.

The most dominant sources of losses in an electrochemical cell, however, are the ohmic losses due to the *internal resistance*. They cause a voltage drop during discharge and are often referred to as the *ohmic polarisation*, or *IR drop*, and are directly proportional to the discharge current applied.

The internal resistance is the sum of several resistances, e.g. ionic resistance in the electrolyte, electronic resistance in the active electrode materials, and interface resistances between current collectors/electrodes, electrode/electrolyte or within the separator. All these resistances are ohmic by nature and dependent on the applied current.

The practical cell voltage can now, in contrast to equations (1.1) and (1.10), with polarisation effects included, be described as:

$$E_{cell} = E_{pos} - E_{neg} - [\eta_{ct} + \eta_c]_{pos} - [\eta_{ct} + \eta_c]_{neg} - iR_{cell} \qquad (1.27)$$

where i is the current, and R_{cell} is the internal resistance of the cell. If a very low current is applied, the polarisations and the *IR drop* are small, or negligible, and thus the cell will operate close to the OCV and most of the theoretical energy is available. Figure 1.23 illustrates the cell voltage decrease as a function of the current used during discharge. Moreover, all polarisation losses are affected by the cell usage and depend on degradation processes; therefore, the complete picture is very difficult to predict.

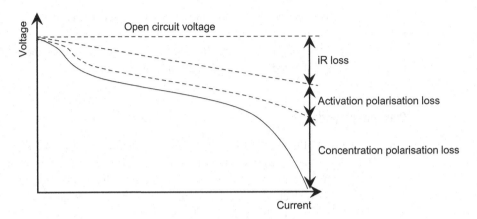

Figure 1.23 The cell voltage as a function of the current used.

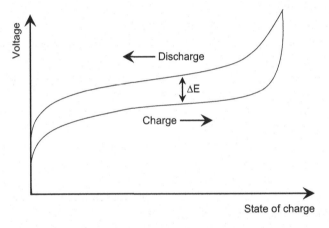

Figure 1.24 The hysteresis effect of a discharge-charge profile.

1.6.1.1   Voltage hysteresis

So far, only discharge profiles have been considered. The reason is to have the fully charged state as a reference for all further discussions. Ideally, the charge behaviour would equal the discharge behaviour, but the kinetics and diffusions of the active species can, and often do, differ between the charged and discharged state. During charging and discharging processes, the chemical and electrochemical reactions are delayed compared to the external load/charger reactions, observed as a voltage difference. This is partially due to the side reactions taking place. In addition, all processes have different time scales to complete the charging or discharging reactions, depending on the chemical constraints. None of these processes is linear or equal with respect to the charging and discharging as the cell has a limit in charge acceptance rate set by the chemical constraints. This may result in observable *voltage hysteresis*, ΔE, in the cell discharge profile, as illustrated in Figure 1.24. Unfortunately, energy, or capacity, is lost due to the lower voltage obtained by the hysteresis. The ΔE may increase with the

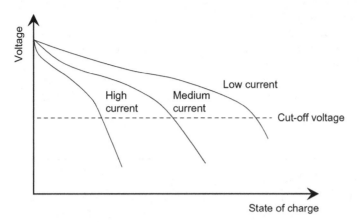

Figure 1.25 Schematic illustration of how the voltage profile is affected by different discharge rates.

charge and discharge rate and has to be considered in the design process both of the battery and the vehicle. In addition, the power electronics must be able to handle the different voltage levels of the charge and discharge processes occurring during vehicle operation, and the control strategies have to be adopted accordingly.

In addition, the cell has an operational voltage range between a fully charged state and a fully discharged state, and is limited by the *cut-off voltages*, i.e. the cell should not be charged or discharged above or below these limits. If these limits are exceeded, the cell is *overcharged* or *overdischarged*.

1.6.2 Charge and discharge rates

Since the cell voltage depends on the current levels used, the cell behaviour will be affected accordingly. The effect of different discharge rates, i.e. the current related to the rated cell capacity,[1] on cell voltage as a function of the charge delivered is illustrated in Figure 1.25. At high discharge rates, much less charge has been delivered before the discharge cut-off voltage is reached, due to large polarisations.

The rate at which the discharge and charge take place is defined as the *C-rate*, expressing the current capability of an electrochemical cell or the constant current charge or discharge rate the cell can sustain for a specific time. A cell discharged at a 1C rate will deliver its rated capacity during one hour. If a cell has a rated capacity of 3 Ah, the 1C rate corresponds to a discharge current of 3 A, and discharging a 20 Ah cell during the same time requires a current of 20 A. The C-rate should not ever be seen as the maximum current a specific cell could deliver.

At the same time, the charging and discharging currents are generally expressed as multiples of C. The discharge time of a cell is inversely proportional to the discharge rate, i.e. nC is a charge or discharge rate that is n times the rated current capacity of the cell. A charge rate of 4C corresponds to a charging time of 15 minutes, while a rate of

---

[1] The rated capacity is often measured as the 1C discharge capacity.

Table 1.1 Discharge currents for some cells having different rated capacity

Rated cell capacity (Ah)	Discharge current for 1C rate (A)	Discharge current for 5C rate (A)	Discharge current for C/2 rate (A)	Discharge current for C/5 rate (A)
3	3	15	1.5	0.6
8	8	40	4	1.6
15	15	75	7.5	3
20	20	100	10	4

0.2C, or C/5 corresponds to five hours of charging. For the same cells as above, discharging the 3 Ah cell with a 4C rate is thus performed by applying a current/load of 12 A, and discharging a 20 Ah cell with a C/5 rate corresponds to applying a current/load of 4 A. All this is further illustrated in Table 1.1 showing the corresponding discharge currents for cells of various capacity and different discharge rates.

Generally, the capacity of an electrochemical cell decreases with increasing discharge currents, since the cut-off voltage is reached faster. Thus, it is crucial to know the discharge conditions when comparing different cells for a certain application. Discharging at 1C rate normally leads to discharge times shorter than one hour. This is mainly due to the non-ideal conditions of the cells. For example, the temperature and the age of the cell may induce side reactions affecting the discharge procedure. Therefore, in some operational modes the *hourly rate* used refers to the current at which the cell will be discharged during a specific time period or number of hours. For applications relying on power output, a *constant power discharge rate* may be applicable and can be expressed analogous to the C-rate described above.

As the voltage under load is different to the equilibrium potential, the capacity varies with the load conditions, and the practical capacity is defined as the current density passing through the cell until the cut-off voltage has been reached. When increasing the current, or C-rate, the cut-off voltage is reached faster, and hence a lower capacity will be achieved. Figure 1.26 illustrates the usable capacity for different discharge rates.

The same pattern, or behaviour, is valid for any charge procedure of value for control strategies for electric vehicles. If the charging current is too high, the upper cut-off voltage, i.e. the highest acceptable voltage of a cell for safety and durability, is reached before the cells are fully charged. Therefore, it is crucial to understand how the C-rates affect cell performance in order to fully utilise the cell capacity, and this is one of the key design parameters when designing a battery for electric vehicles. The charge/discharge behaviour is highly influenced by the cell temperature. For example, fast charging of a Li-ion cell at low temperature, for instance below 10° C, can be detrimental for the durability and safety of the cell.

1.6.3   Capacity

The total charge, Q, is the charge transported by the current during a specific time period and Q/weight (mainly cell weight is used) shows the *capacity* of the cell. The latter is

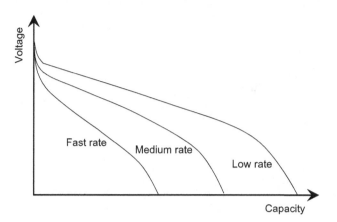

Figure 1.26 Capacity variations depending on the discharge rates.

related to the amount of active electrode material in the cell. A small cell has less capacity than a larger cell with the same chemistry, but the OCV is the same.

The theoretical capacity of an *electrode* is related to the charge transferred in a specific reaction. It is often given as *specific capacity* (Ah/kg or mAh/g) or *volumetric capacity* (Ah/L) and it is derived from Faraday's law:

$$Q_{th} = \frac{nF}{M_w} \tag{1.28}$$

where Q_{th} is the theoretical specific capacity in Ah/kg or mAh/g, n is the number of transferred electrons, F is Faraday's constant (96485 As/mol) and M_w is the molecular weight of the electrochemical active material (in g/mol). It should be noted that the theoretical capacity is usually given according to the discharged state of the active material. The fraction of the stored charge an electrochemical cell can deliver depends on several factors, including chemistry, discharge rate, voltage level, and temperature. For example, when discharging at rates that are too high, the cut-off voltage will be reached before the full capacity has been delivered (Figure 1.25).

To illustrate how the capacity relates to the discharge rate, the *Peukert law* describes the delivered capacity of an electrochemical cell in terms of the discharge rate. At increased rates, or current densities, the available capacity is reduced, resulting in reduced utilisation of the cell. The Peukert law is an empirical formula and can be written as:

$$C = i^k t \tag{1.29}$$

where C is the capacity at a one-hour discharge rate (i.e. 1C) (in Ah), i is the discharge current, t is the discharge time, and k is the Peukert constant. The value of the Peukert constant, k, varies with the cell chemistry, as well as with the internal resistance of the cell. Higher discharge currents (C-rates) result in larger losses and less available capacity. This can be further illustrated by plotting the capacity reached at the cut-off voltage as a function of the discharge rate in a log-log plot, a (modified) Peukert plot, as is shown in Figure 1.27. Note the very fast decrease in capacity at high C-rates, which is a common feature of most electrochemical cells.

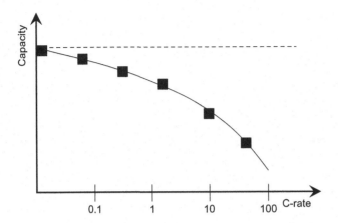

Figure 1.27 A modified Peukert plot. The dotted line represents the maximum available capacity at the cut-off voltage.

1.6.4   Energy and power

The driver of an electric vehicle has an interest in knowing how far and how fast the vehicle can go. By definition, the energy of an electrochemical cell refers to the amount of useful work the cell can perform until the cut-off voltage has been reached, whereas the power of the cell refers to the rate at which this work can be performed; i.e. basically the distance and speed of the vehicle.

The energy of an electrochemical cell is defined as:

$$Energy = \int_{0}^{t_{cut-off}} V(t)Idt \tag{1.30}$$

and thus corresponds to the area under the discharge profile (e.g. Figure 1.26). Often the maximum available capacity is given according to the rated cell voltage in order to report the energy content of a cell. This, however, gives a misleading picture of the *available* energy content of the cell if only operating at the narrow state of charge ranges, as in a hybrid electric vehicle (Section 5.1).

The *specific energy*, i.e. the energy per weight unit [Wh/kg], of an electrochemical cell can be related to the active electrode material only (see above) or to both the active and the non-active materials in the cell. The energy density is dependent on cell voltage and capacity; there are some general routes to be considered to increase these parameters: (i) increase the average electrode potential, (ii) increase the number of electrons exchanged per transition metal, and (iii) decrease the molecular weight per mole electrons exchanged. According to equation (1.28), the theoretical capacity is proportional to the number of electrons involved in the redox reactions, and inversely to the molecular weight of the active material.

The kinetics of the electrode material is, however, the key for utilising as much as possible of the theoretical capacity. The capacity of each electrode can be determined separately in half cell measurements. The power of an electrochemical cell, normally given as [W/kg], is defined as the energy delivered during a specific time according to:

$$Power = \frac{\int_0^{t_{cut-off}} V(t)Idt}{\int_0^{t_{cut-off}} dt} \quad (1.31)$$

How quickly the energy can be delivered from a cell depends on the kinetics of the electrodes and the electrolyte, especially at the interfaces between them. Therefore, it is not possible to get the same energy delivered at any time interval. For example, even if 10 kWh can be delivered from a battery of an electric vehicle during 10 seconds, 20 kWh can most likely not be delivered from the same battery during 5 seconds, despite being the same amount of energy per time unit, since the kinetics will influence the power capability. It should therefore be carefully noted that the time to reach the cut-off voltage in equation (1.31) is not necessarily the same as in equation (1.30).

Increasing the energy of an electrochemical cell is accomplished by either an increase in the voltage or in the specific capacity, e.g. by increasing the amount of the active materials. Enhancing the power of an electrochemical cell, however, requires a reduction of the voltage drop with increased current, which can be achieved by decreasing the ohmic, kinetic, and mass transfer limitations of the cell. The most common example is to use thinner electrodes to reduce diffusion limitations, hence less active material. As a consequence of these counter-acting requirements for energy and power, there is always a trade-off between the two: a cell can either be *energy-optimised* or *power-optimised*. High-power output from an energy-optimised cell will result in large polarisation and consequently large losses. It is therefore important to use a cell optimised for the intended use; power-optimised cells for hybrid electric vehicles are far from suitable for all-electric vehicles, and vice versa.

A *Ragone plot* is a useful tool to characterise the trade-off between energy and power for electrochemical cells differing in chemistries, shapes, sizes, and weights. The available energy density as a function of the available power density on a logarithmic scale is used for illustration, as shown in Figure 1.28. The supporting diagonal lines in the Ragone plot indicate the relative discharge (or charge) time, i.e. the C-rates, and the Ragone plot displays how much energy and power is available for each operational mode, e.g. acceleration of a vehicle. The relationship between the power and energy ability of a given cell is given by the curve in the Ragone plot. In case of fast operational processes, i.e. at high C-rates (A in Figure 1.28), high-power can be achieved, but the delivered energy is very limited. On the other hand, fairly high specific energy (B in Figure 1.28) can be delivered during longer time periods with a very low specific power output, and for an electric vehicle these constraints limit the acceleration performance,

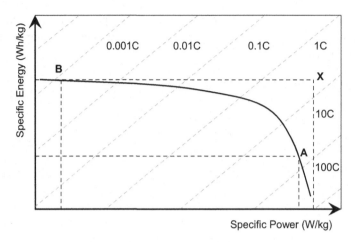

Figure 1.28 Ragone plot.

as well as the all-electric driving range. Consequently, the maximum power can only be utilised with an infinitesimally small energy, and vice versa.

Thus, it is impossible to simultaneously achieve both maximum energy and maximum power density (X in Figure 1.28). Therefore, these properties must be jointly optimised depending on the application in mind. In the Peukert diagram (Figure 1.27), capacity utilisation of the cell and the corresponding capacity losses are given. In the Ragone plot, on the other hand, both the loss in capacity utilisation and the voltage losses are given. The Ragone plot is a common tool for understanding different chemistries and provides useful information when comparing various technologies, information such as limitations in available power and optimal operational region; meaning the region where both the power and energy densities are high. Therefore, the Ragone plot is much more used than the (modified) Peukert plot during an electric vehicle design process, in order to determine the cell applicable to different types of vehicles.

1.6.5   Efficiency

In an ideal electrochemical cell, the ionic current through the electrolyte inside the cell is equivalent to the electronic current through the external load. However, the performance and the efficiency of a cell depend both on charge and mass transfer. The magnitudes of these parameters are in turn affected by diffusion rates, ion conductivity, as well as the stability of all materials within the operating voltage range. Often two efficiency metrics are used for electrochemical cells: *Coulombic efficiency* and *energy efficiency*.

The Coulombic efficiency is defined as the charge transferred during the discharging process, Q_{dis}, over the charge transferred during the subsequent charging process, Q_{cha}:

$$\eta_{Coulombic} = \frac{Q_{dis}}{Q_{cha}} = \frac{\int\limits_{dis} Idt}{\int\limits_{cha} Idt} \qquad (1.32)$$

Energy efficiency is defined similarly, but as the energy transferred according to:

$$\eta_{energy} = \frac{\int\limits_{dis} V(t)Idt}{\int\limits_{cha} V(t)Idt} \qquad (1.33)$$

To be practically interesting, cells for electric vehicles must have both these efficiencies close to unity – values of 0.99 are not uncommon under optimal usage conditions.

1.7 Electrochemical analysis methods

The properties and performance characteristics of any electrochemical cell need verification and there are a large number of analysis methods that can be applied. Some parameters of different origins may be difficult to separate with the necessary accuracy by simple methods, such as just applying potential steps to record the current-time dependency. Therefore, a combination of methods is often required. Overall, the electrochemical behaviour can be studied using several methods, often divided into steady state and dynamic techniques. Below some fundamental and frequently used methods are described: galvanostatic and potentiostatic cycling, cyclic voltammetry, impedance spectroscopy, and the use of a reference electrode.

1.7.1 Galvanostatic and potentiostatic cycling

Galvanostatic and *potentiostatic* cycling methods are the test methods most commonly used to study the performance of electrochemical cells for electric vehicle applications. The kinetics and the electrode mechanisms can be studied by galvanostatic cycling, and mass-transfer and diffusion properties can be investigated using potentiostatic cycling.

In the case of galvanostatic cycling, a constant current is applied, often at a specific C-rate, and the voltage response is measured. This method results in voltage profiles clearly showing the changes in the electrochemical performance during cycling. The method can be used to qualitatively compare cells and determine the cell capacity. The galvanostatic cycling method is often used to follow the degradation processes occurring as a consequence of cell usage, and will hence be further discussed in Section 7.3.1.

Potentiostatic cycling, on the other hand, involves holding the cell potential, or cell voltage, constant and follows the decline of the current. As the redox reactions proceed, the current decreases approaching zero when the redox reactions are completed.

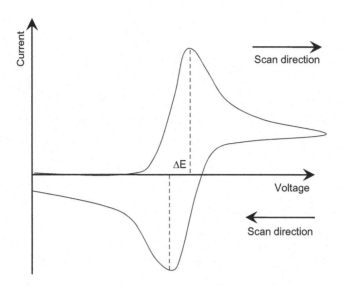

Figure 1.29 Cyclic voltammogram of a single redox couple.

Experimentally the constant potential is applied until equilibrium is reached, or until the current level has declined below a predefined level. Thereafter, the next potential is set and the procedure is repeated.

1.7.2   Cyclic voltammetry

One commonly employed analysis method is *cyclic voltammetry*, which is used to obtain information about the different electrode reactions. The method is based on linearly changing the voltage (E) of the cell at equilibrium according to:

$$E = E_i \pm v_j t \tag{1.34}$$

where E_i is the initial potential and v_j is the scan rate (V/s). The measured current is then plotted versus the applied voltage resulting in a wave-formed curve as schematically shown in Figure 1.29. As the predefined cut-off voltage is reached, which depends on the cell chemistry, the scan direction is reversed and usually the scan rate is the same in both directions (but can be altered if needed). This procedure can be repeated several times during a single experiment to observe changes affected by, for example, voltage ranges and scan rates.

The current peaks appearing in the voltammogram correspond to redox reactions taking place in the electrochemical cell (within the measured voltage range). The current will increase as the voltage reaches the redox potential of the species, and then fall off as the concentration of the species is depleted. If the electrochemical reaction is reversible a current peak will be obtained in the other scan direction as well. The two peaks usually have similar shapes, but they often differ slightly in voltage, ΔE, giving information about both the redox potentials and electrochemical reaction rates. The main reason is that the reversible redox couples exhibit polarisation overpotentials and thus ΔE is a

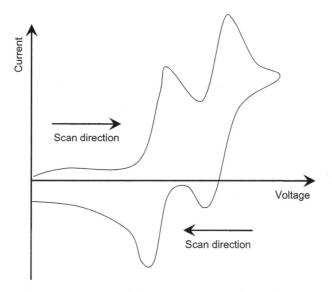

Figure 1.30 Cyclic voltammogram involving at least two redox reactions.

hysteresis between the reduction and oxidation peaks, as can be seen in Figure 1.29. The hysteresis emerges from diffusion limitations and intrinsic activation barriers of electron and/or charge transfer and corresponds to the same hysteresis effect as described in Section 1.6.1.1. The shape of and the distance between the peaks thus depend on the reversibility of the redox reactions. Reversible, diffusion-controlled reactions will exhibit an almost symmetrical pair of peaks with a hysteresis independent of the scan rate according to:

$$\Delta E = \frac{2.3RT}{nF} \tag{1.35}$$

In the case of quasi-reversible reactions, the peaks are more separated, and have non-uniform shapes. Completely irreversible reactions exhibit only a single peak observed in either scan direction. As more complex electrode processes occur and for electrode materials having two or more redox species, the cyclic voltammogram can exhibit several peaks during reduction and oxidation scans, respectively. This is illustrated in Figure 1.30, and for these systems it is difficult to separate out the different redox reactions and relate them to specific electrochemical species or processes within the active electrode materials.

In order to study the redox reaction for each electrode independently, half cell measurements are commonly used.

1.7.3   Electrochemical impedance spectroscopy

The dynamic characteristics of an electrochemical cell are valuable for the control system in an electric vehicle, where cell performance characteristics must be known at each stage in order to fulfil the vehicle driving conditions. These dynamic properties can be achieved by monitoring the frequency response. *Electrochemical impedance spectroscopy* (EIS) is a direct and non-destructive method to study processes by measuring the change in the *electrical impedance, Z*. A single sinusoidally varying signal is applied to the electrochemical cell at open circuit voltage and the phase-shift and amplitude of the resulting signal are measured. The response can be described as the impedance and is measured over a wide frequency range, normally MHz to mHz. EIS can be used to observe changes in mass transport, double-layer capacitance, and ohmic resistance in the electrochemical cell, as well as the electrodes uniquely.

The potential and current can be described according to:

$$E_{cell} = E^0 \sin(\omega t) \text{ and } i = i^0 \sin(\omega t + \phi) \tag{1.36}$$

The impedance as a function of the frequency can be represented by:

$$Z(\omega) = \frac{E}{i} = Z^0 \exp(\sqrt{-1}\phi) = Z^0(\cos\phi + \sin\phi) \tag{1.37}$$

The frequency dependent impedance, $Z(\omega)$, can be described in an imaginary part, Z_i, and a real part, Z_r. In a Nyquist plot (Figure 1.31), Z_i is plotted versus Z_r, and can be interpreted in terms of different parameters originating from electrode and electrolyte interactions.

The different regions in the Nyquist plot correspond to different processes: the resistance between different cell components, kinetics and capacitive features, as well as electrolyte and solid phase diffusions. The intercept of the real part at high frequencies refers to the bulk electrolyte resistance, R_s, and the charge-transfer resistance, R_{ct}, refers to the local minimum at lower frequencies (R_s+R_{ct} in Figure 1.31). The double-layer capacitance, C_{dl}, is related to the local maximum ω_{max} of the semi-circle according to:

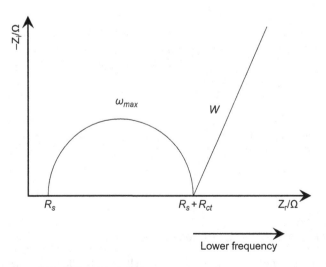

Figure 1.31 A Nyquist plot.

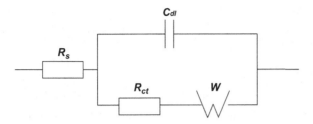

Figure 1.32 An equivalent circuit corresponding to the Nyquist plot in Figure 1.31.

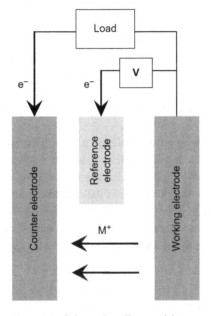

Figure 1.33 Schematic cell comprising a working electrode, a counter electrode, and a reference electrode.

$$\omega_{max} = \frac{1}{R_{ct}C_{dl}}$$ (1.38)

The tail of the Nyquist plot at low frequencies refers to the diffusion controlled electrochemical processes, in both the electrolyte and in the porous electrodes. The diffusion is often represented by the Warburg impedance, W, and can be both of infinite and finite character, depending on the diffusion layer thickness and the diffusion coefficient associated with the R_{ct} and the C_{dl}.

The semi-circle obtained in the Nyquist plot represents different parameters, which can be empirically modelled as an equivalent circuit consisting of several elements. Figure 1.32 shows one possible equivalent circuit for the Nyquist plot in Figure 1.31.

Depending on the electrochemical system selected and the age of the cell, there can be two or more semi-circles in the Nyquist plot representing, for example, a film that has been formed on the electrode. Thus the EIS technique can also be used to study how different cells age. Depending on the nature of the obtained spectra, for new and old cells respectively, information about the degradation processes can be acquired.

1.7.4   Reference electrode

The common two-electrode set-up with a negative and a positive electrode is used to study the overall electrochemical cell properties. In order to study the individual electrodes and their corresponding properties, a three-electrode set-up can be used as a complement to half-cell measurements. This set-up is shown in Figure 1.33 and consists of: the *working electrode* (W.E.), or the electrode of interest, the *counter electrode* (C.E.), and the *reference electrode* (R.E.).

As the name indicates, the reference electrode is used to provide an electrode potential to which the other electrode potentials can refer. The reference electrode must therefore possess a stable potential regarding time and temperature, and should be unaffected by the main electrochemical reactions of interest. The selection of reference electrode, as well as the position relative to the working electrode surface, may sometimes affect the results.

During cycling, the performance of the negative and the positive electrodes can be studied separately, and the use of a reference electrode is therefore a powerful tool in order to understand the different electrode processes and the interactions between the working and the counter electrodes (i.e. the positive and the negative electrode). It is, however, a useful tool only in the cell development phase or for the understanding of aged cells disassembled for post-mortem analysis (Section 7.3.6) since it is in most cases impossible to introduce reference electrodes in commercial cells applicable for electric vehicles.

2 Battery technologies for electric vehicles

Now it is time to take the next step towards a battery for an electric vehicle. The fundamental electrochemical processes in the previous chapter will now be dressed in 'chemistry' and materials. Due to the combinations available for electrodes and electrolytes, a wide range of battery technologies can be obtained, all having specific cell properties. The limitations and possibilities of the different combinations will affect cell performance considerably, but not all of them are of interest for electric vehicles. In the following chapter, only those concerning vehicle applications will be discussed and only for rechargeable battery technologies having acceptable energy and power capabilities for the demanding vehicle requirements. These technologies will be described in general and schematic terms, which may also be of interest for other types of applications, such as stationary and smart grid applications.

By combining a positive and a negative electrode in different ways, and then immersing them in a suitable electrolyte, a palette of cell potentials, charge/discharge characteristics, capacity, power and energy densities, cycle life, for example, can be obtained. Depending on the intended usage, some technologies and material combinations are more suitable than others. Indeed, the performance of a cell can vary significantly during actual operational conditions compared to ideal laboratory tests, and therefore the performance conditions should be evaluated. Many of the factors influencing cell performance under different operation conditions are linked together. The same battery technology may differ significantly between manufacturers and manufacturing processes due to specific cell constraints, e.g. material specifications and cell design. The cells may also vary from batch to batch from the same manufacturer. Moreover, storage and the age of the cells influence performance, which can change drastically; the performance obtained at *beginning of life* (BOL) is rarely the same as at *end of life* (EOL).

The following battery technologies will be described:

- lead-acid (Pb-acid)
- nickel metal-hydride (NiMH)
- lithium
- high-temperature molten-salt
- nickel-zinc (NiZn)
- zinc-air (Zn-air)
- metal-ion (Me-ion)
- redox flow

In addition, high-energy density capacitors will also be described. They are not batteries in the classical sense, but still an electrical energy storage technology of considerable interest for electric vehicles. Another suitable technology for electric vehicles is fuel cells. It is an energy conversion technology suitable either for primary propulsion of all-electric vehicles or for powering auxiliaries. As fuel cells often are less suitable for fast and transient power responses, a battery usually complements them in a vehicle.

2.1   Lead-acid batteries

The lead-acid battery, Pb-acid, is the most widely used battery technology in the automotive industry; employed as the power supply for the starter motors and as voltage regulators during vehicle use. Pb-acid batteries come in almost any size and voltage level, and the main reason for their wide usage is the ability to supply high surge currents.

With more than 140 years of development, the Pb-acid battery is a reliable and robust technology. It is due to a low internal impedance that Pb-acid batteries can deliver very high currents. Since lead is a heavy metal (11.3 kg/dm^3) though, the technology itself becomes heavy and bulky and due to their environmental impact, lead and other heavy metals are more or less prohibited to be used in many applications. This, however, has resulted in Pb-acid batteries being one of the most recycled products in the world. In conclusion, the weight and the power capabilities of this technology make it best suited for electric vehicles of micro-hybrid configurations, i.e. start-stop functionality, which will be described more in Section 5.1.

2.1.1   Basics

The active materials in the Pb-acid battery are lead dioxide (PbO_2) for the positive electrode and metallic lead (Pb) for the negative electrode; both electrodes being of blocking electrode character (Figure 1.3). Sulphuric acid is used as the electrolyte. During discharge of the cell, Pb is oxidised and PbO_2 is reduced, both electrodes will consist of $PbSO_4$ in the discharged state, and the sulphuric acid concentration in the electrolyte decreases as water is produced, all according to the following reactions:

$$\text{Oxidation reaction: } Pb + HSO_4^- \rightarrow PbSO_4 + H^+ + 2\ e^-$$

$$\text{Reduction reaction: } PbO_2 + HSO_4^- + 3H^+ + 2e^- \rightarrow PbSO_4 + 2H_2O$$

$$\text{Overall reaction: } PbO_2 + Pb + 2H_2SO_4 \rightarrow 2PbSO_4 + 2H_2O$$

The redox reactions are reversible and the reverse reactions take place during the subsequent charging sequence. A Pb-acid cell has a nominal cell voltage of 2.0 V, and a typical voltage profile vs. discharge time is given in Figure 2.1.

There are two forms of PbO_2; the orthorhombic columbite-structured α-PbO_2 and the tetragonal rutile-structured β-PbO_2. The β form is the most electrochemically active and has a higher capacity per unit, but a shorter cycle life than the α form. In a commercial

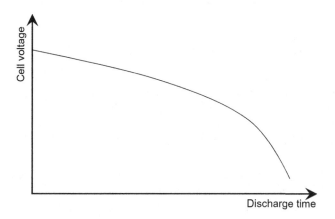

Figure 2.1 A typical voltage profile for a Pb-acid battery as a function of state of charge.

Pb-acid battery, the positive electrode consists of the α and the β forms due to the electrode production procedures.

Because lead is a soft metal, alloying is needed to obtain an electrode material manageable in cell production, and here antimony (Sb) provides harder electrodes. Sb is, however, dissolved during the charge/discharge procedures resulting in capacity loss. At low concentrations of Sb (<5%), small amounts of other alloying elements are needed in order to minimise the grains and the defects in the electrode, as well as to stiffen the electrodes. These elements are usually one or more of the following: sulphur (S), copper (Cu), arsenic (As), selenium (Se), tellurium (Te), or alkaline earth elements. By changing the electrode composition, the Pb-acid batteries can be tailored for different applications and usage.

The electrolyte, an aqueous solution of sulphuric acid (H_2SO_4), plays an active part in the charging/discharging reactions (see above) and its concentration is essential in terms of cell performance. As mentioned above, at a fully charged state the cell voltage is about 2.0 V, and it is a function of the electrolyte concentration according to the Nernst equation (equation (1.9)):

$$E = 2.047 + \frac{RT}{F} \ln\left(\frac{aH_2SO_4}{aH_2O}\right) \tag{2.1}$$

Aside from heavy weight, usage at low SOC levels and insufficient charge acceptance are the main disadvantages of the Pb-acid technology. The latter is due to limitations in the kinetics of the electrochemical reactions. The benefits of the technology are found at high SOC levels, but this would result in a largely oversized battery for all-electric vehicles where cycling over a wide SOC range is preferred for an acceptable driving range.

In Pb-acid cells, many of the wanted, and unwanted, reactions have low activation energies and can be spontaneous. From a durability perspective, the electrolyte should preferably be added just before use, resulting in a technology having an indefinite shelf life when stored without the electrolyte. As soon as the electrolyte has been added, however, Pb-acid batteries must be stored in a charged state to avoid deterioration of the active materials.

The dominating side reactions for Pb-acid cells are *self-discharge, sulphation, stratification* or gassing of the electrolyte, and *shedding*. These reactions will all hamper performance over time and are explained one by one below.

In terms of self-discharge mechanisms, both Pb and PbO_2 are thermodynamically unstable in the H_2SO_4 electrolyte, and will form $PbSO_4$ according to the same reactions as those taking place during normal discharge conditions, and consequently the self-discharge rates are dependent on the storage temperature, as well as the design and composition of the electrodes. For example, low-antimony or antimony-free alloys have slower self-discharge rates. The amount of antimony, however, has to be balanced in order to optimise other performance and mechanical parameters of the cell.

The self-discharge of the cell will have a considerable effect on the sulphation of the cell. During the sulphation process, the cell loses the ability to be charged under normal charging conditions. At the discharged state, both electrodes comprise $PbSO_4$ as crystallites on the surface of the electrodes-sulphation. The first formed $PbSO_4$ is amorphous and appears as a homogeneous layer covering the electrode surface. The reversible reaction back to Pb, PbO_2, and H_2SO_4 is possible if the amorphous phase is present during the battery recharge process. As the battery is repeatedly cycled, the reversibility of $PbSO_4$ is not complete and the amorphous phase slowly converts to a stable crystalline form that no longer dissolves on recharging.

The crystallisation of $PbSO_4$ proceeds on the surface of the Pb electrode and physically minimises the contact area between the electrode and the electrolyte, thus blocking further redox reactions. As the available surface area of the electrodes is reduced, the internal resistance increases, and hence the capacity of the cell irrevocably decreases. The capacity loss is irreversible due to the loss of usable active material over time. The limited active surface area also affects the charging conditions, resulting in extended charging time, less efficient and incomplete charging, and higher cell temperatures. The $PbSO_4$ crystals can easily grow when stored during prolonged periods and at low states of charge. In addition, elevated temperatures enhance the reaction rate.

The sulphation reaction may also occur if the H_2SO_4 concentration in the electrolyte is higher than normal, e.g. due to loss of water during overcharging. Consequently, Pb-acid cells must be fully charged occasionally, and the water content must be controlled. The sulphation reaction occurs in all Pb-acid cells, but can be limited by controlled charging conditions. It is possible to overcome sulphation by *desulphation* and to convert the $PbSO_4$ back to Pb and PbO_2 by fully charging the battery at low currents, or by high current pulses for very short time periods. The low charging rate will bring the chemical reactions close to equilibrium and a pulse-charging procedure will break down the $PbSO_4$ crystals in order to liberate the electrode surface for further electrochemical reactions. During charging, the electrodes may, however, crack, and in extreme cases, depending on the cell design, this may cause an internal short circuit of the cell due to contact between the electrodes.

Concentration variations, stratification, of the electrolyte, as a result of differences in density, temperature variations within the cell, and storage conditions, can cause a separation of the electrolyte into regions more diluted or containing a higher concentration of H_2SO_4 in the bottom of the cell. This stratification results in concentration variations across the electrode surfaces and thereby uneven utilisation of the electrode. During the charging process, and at potentials above 2.4 V, the electrolyte can generate hydrogen and oxygen gases due to electrolysis of the water in the electrolyte. The gassing may cause an explosive atmosphere and abuse situations. By adjusting the voltage level, the electrolysis can be hampered. It is important to remember that there are usually differences between a fresh and a used Pb-acid cell in terms of charging procedures due to degradation processes taking place during the life of the battery.

Shedding, or exfoliation of lead from the electrodes, can occur as a consequence of insufficient electrode construction and cell design. This normally originates from high charging rates or demanding cycling behaviours. Exfoliated lead will accumulate in the electrolyte compartment and result in severe damage to the electrode, irreversible capacity loss, and possibly a short circuit. There are design options and operational restrictions to overcome some of the shedding effects, e.g. the structure of the active material, a limited number of cycles and charge states, and specific charging methods.

2.1.2   Lead-acid concepts

Lead-acid batteries have been developed and tailored for different applications and needs. To improve performance characteristics such as charge acceptance and to inhibit sulphation and other side reactions, changes primarily in the electrode compositions have been made. In the following sections, a brief review of the most common cell concepts is given: high-power and high-energy cells, and sealed cells.

2.1.2.1   High-power and high-energy cells

As for all electrochemical cells, the Pb-acid cells can be designed for either power or energy. Power is an essential parameter for many vehicle applications. At the cranking of a vehicle, the battery is fully charged and only a few percent (*ca.* 2–5%) of the charge capacity is used for the cranking procedure, and thereafter the alternator recharges the battery. There are Pb-acid batteries especially designed for this purpose: the *starting, lighting*, and *ignition* batteries (SLI).

Low discharge states cause severe damage to the electrodes with a considerable negative effect on durability. The reason for this is a thin-electrode design with a large surface area in order to achieve high current capability, which is beneficial in terms of specific power and reduced internal resistance. The SLI battery will show long life (if not deeply discharged) due to the frequent usage of pulse-like characteristics limiting, for example, the sulphation reactions. The SLI battery is the most commonly used Pb-acid battery in non-electric vehicle applications, i.e. conventional vehicles.

On the other hand, energy is the key performance parameter in many applications, such as electric vehicles, marine applications, and forklifts. Deep, or complete, discharge cycles reaching low SOC levels are therefore needed in order to design a battery of reasonable size. Deep discharge cycles of a Pb-acid cell normally result in a temperature increase due to the high internal resistance. Thicker and reinforced electrodes are used to minimise heat evolution. Thick electrodes, however, result in a lower charging capability, and therefore the battery has to be oversized if used in applications requiring some fast charging. Another route is to combine a Pb-acid battery with an asymmetric capacitor (Section 2.4.2) in a single cell, where the Pb-acid part is responsible for fulfilling the energy requirements and the capacitor part fulfils the power requirements. This cell design is an attractive alternative for electric vehicles, especially for hybrid electric vehicles, but only when the physical size is of less importance.

2.1.2.2   Sealed cells

To overcome the deficiencies of the electrolyte, i.e. the gassing or evaporation, sealed Pb-acid cell designs have been developed. There are at least three types of sealed designs: *valve regulated* (VRLA), *absorbed glass mat* (AGM), and *gel* batteries.

The VRLA battery has a pressure valve, which opens in abuse situations. Handling and maintenance are favourable features of VRLA batteries due to the sealed and leakage-free design. Most Pb-acid batteries are designed to avoid free electrolyte, and this is commonly achieved by using a separator soaked in electrolyte. In the AGM type battery design, a separator made of a boron silicate ($BSiO_4$) fibreglass mat is used to absorb the electrolyte and at the same time promote the recombination of H_2 and O_2 into water. This recombination can be further enhanced by electrolyte additives, limiting long-run degradation inside the cell. The absorbed electrolyte in the mat will enhance electrolyte concentration near the electrode surfaces, allowing higher charge/discharge rates, and it is also favourable for deep cycling. The self-discharge rate of AGM cells is slow due to limited excess of electrolyte. The mat also acts as a reinforcement of the cell, resulting in a design able to better withstand vibrations. Another route to minimise the excess of electrolyte is to use a gel-like electrolyte where the gel acts as both electrolyte and separator. Here, the charge rate is limited in order to prevent gas evolution and decomposition of the gel, which would damage the cell. Both the gel-type and AGM cells can also be designed as VRLA.

2.2   Nickel metal-hydride batteries

The nickel metal-hydride (NiMH) battery technology was commercialised as late as 1989 for portable computers and cellular phones. During the 1990s, NiMH batteries dominated the cellular phone market, but experienced a rapid decline as the more energy dense Li-ion concept was introduced. NiMH is still widely used in consumer applications and as energy storage systems in hybrid electric vehicles. The technology was developed for environmental reasons, and can be seen as a further development of the nickel cadmium (NiCd) technology, which is now forbidden to be used in vehicle applications due to the heavy metal cadmium content. The materials used in NiMH batteries are, however, environmentally friendly, though many of the metals used are rare.

Normally, NiMH cells have low internal resistance and deep cycling is possible; several hundred thousands of cycles can be achieved at large SOC ranges. The technology is robust and tolerant for fast charge and discharge conditions over a wide range of temperatures. During storage and uncontrolled usage, gas may be emitted (O_2 and/or H_2) and therefore the cell design must incorporate safety vents for protection.

2.2.1   Basics

The positive electrode consists of nickel oxyhydroxide (NiOOH), and the negative electrode is a metal alloy (*M*) able to absorb hydrogen, the active element, during charging. Careful selection of the alloy constituents and proportions allows balancing of the thermodynamics to enhance the absorption and release processes of hydrogen. During discharging of the NiMH cell, protons (H^+) are extracted from the metal alloy *M* by reaction with hydroxide (OH^-) of the electrolyte to form water, and incorporated in the Ni-based electrode according to the following reversible electrode reactions:

$$\text{Negative electrode: } MH + OH^- \rightarrow M + H_2O + e^-$$

$$\text{Positive electrode: } NiOOH + H_2O + e^- \rightarrow Ni(OH)_2 + OH^-$$

$$\text{Net reaction: } NiOOH + MH \rightarrow Ni(OH)_2 + M$$

The standard potentials for the negative and the positive electrodes are +0.52 V and –0.83, respectively, resulting in an OCV of 1.35 V and a rated nominal cell potential of *ca.* 1.2 V. During the subsequent charging, protons are reinserted into the metal alloy according to the reactions shown above, but in reverse. The discharge profile is flat for a large part of the SOC range, as shown in Figure 2.2.

The charge profile, on the other hand, displays a shape other than the discharge profile, and the voltage maximum is observed *after* the fully charged state has been reached, as schematically illustrated in Figure 2.3. During the charging process, the internal cell temperature rises. At a state of charge of about 70–80%, a more rapid increase in the

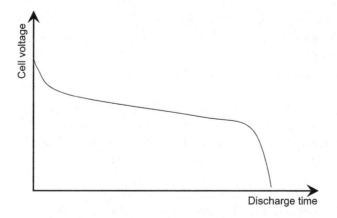

Figure 2.2 Schematic discharge profile of a NiMH cell.

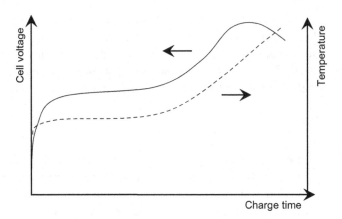

Figure 2.3 Schematic charge profile (solid line) and the associated temperature increase (broken line) of a NiMH cell at 1C rate and 20 °C ambient temperature.

voltage is observed due to the on-set of the oxygen evolution reaction at the positive electrode, an exothermal reaction, and a corresponding temperature increase is observed (Figure 2.3). As can be seen, the temperature also increases slightly in the first phase of the charging process; this is due to heat originating from the internal cell resistance.

The temperature increase causes the voltage to drop slightly as the cell is fully charged and an overcharge procedure dominates. Any overpressure within the cell causes a potential abuse situation, and the charging of a NiMH cell is therefore restricted in order to avoid overcharging. A constant voltage charging procedure cannot be applied. Two different charging procedures are therefore normally used for charging NiMH cells: the *voltage control* and the *temperature control* methods. The voltage control method is based on accurate control of the voltage changes during the charging process. As the voltage increases and the cell reaches the fully charged state, the cell potential drops slightly. This drop can be monitored to interrupt charging. At slow charging rates, however, the voltage drop can be very small, or even absent, limiting the control and hence an abuse situation may be ineluctable.

Similar to the voltage control method, the temperature control charging method also measures temperature changes during the charging process. As the process continues, much of the charging energy is converted to heat, and the change in temperature can be detected using a temperature sensor. The charging procedures will be further described in Section 6.1.1.

If the cell is overcharged, the positive electrode reaches full charge before the negative electrode, and oxygen will evolve at the NiOOH electrode. If passing through the porous separator, oxygen can react with hydrogen at the surface of the negative electrode to form water, *the oxygen recombination reaction*.

$$4MH + O_2 \rightarrow 4M + 2H_2O$$

This reaction will ensure that no internal pressure is built up and can be seen as an internal safeguard. The reaction only takes place if there is an excess of MH electrode material to ensure enough available hydrogen atoms. Thus, as long as there are enough hydrogen atoms available, the cell can be overcharged without damage. By this exothermal reaction, however, the cell is locally heated, affecting the equilibrium pressure of the metal hydride electrode, which can lead to hydrogen evolution and increased pressure, and at a certain point causing an abuse failure. By using an MH electrode with a higher capacity than the NiOOH electrode, it is possible to minimise the overcharge sensitivity.

Normally, the MH/NiOOH ratio, or the *N/P ratio*, is found in the range of 1.3 to 2.0, and the ratio can be used to design cells with different characteristics. Cells optimised for high-power (e.g. hybrid electric vehicles) have a high ratio, and cells optimised for energy (e.g. all-electric vehicles) have a lower ratio. As a consequence, the NiOOH electrode is the limiting electrode, which determines the usable cell capacity. Moreover, as a life-prolonging feature, designing the cell with excess negative electrode capacity will inhibit oxidation/corrosion. The cycle life is affected by the active negative electrode material and its composition, microstructure, and electrode fabrication processes. Moreover, the cell design and electrode stacking, the N/P ratio, the separator, and the purity of the electrolyte must be taken into account.

Charging and discharging of the cell will cause wear and tear on the materials. The absorption and desorption of hydrogen result in expansion and contraction of the positive electrode material, respectively, both of which are likely, at some point, to induce structural defects. As an example, the particle size may be reduced, resulting in increased diffusion polarisation at the surface of the electrode. Additionally, the $Ni(OH)_2$ layer formed on the surface during cycling can be thicker, acting as an inhibitor of the oxygen recombination reaction. This in turn results in an increased amount of generated oxygen, as well as hydrogen, which may result in an increased local electrode temperature and cell pressure build-up.

Moreover, at high charging rates, the amount of oxygen produced may be too large for the reaction kinetics of the recombination reaction, resulting in increased oxygen pressure. The pressure will be released via a safety valve, and, as a consequence, electrolyte may be consumed, resulting, subsequently, in a dry-out of the separator. A complete discharge may result in *polarity reversal*, which can cause irreversible damage. This may occur if cells with slightly different capacity are connected in series where a cell of lower capacity will be completely discharged before the others. The cells of higher capacity will then force the discharged cell into reverse. Therefore, the end-of-discharge voltage of the cells must be monitored. Moreover, polarity reversal may be even more severe if there is a temperature difference between the cells, as the capacity of NiMH cells significantly declines at low temperatures.

2.2.2   NiMH battery materials

The metal alloy used as the negative electrode is actually an intermetallic alloy able to reversibly form hydrides, i.e. that can incorporate/release hydrogen in large volumes – several hundred times its own volume. There are mainly three types of alloys used in the NiMH batteries of today; AB_2, A_2B_7, and AB_5 materials. A is here normally a mix of the rare earth metals lanthanum (La), cerium (Ce), neodymium (Nd), and/or praseo-dymium (Pr), and B is usually nickel (Ni), cobalt (Co), manganese (Mn), and/or aluminium (Al). For high capacity cells, the AB_2 material is made differently; A is titanium (Ti) and/or vanadium (V) and B is zirconium (Zr) or Ni, modified with chromium (Cr), Co, iron (Fe), and/or Mn.

The AB_5 alloys are widely used, even if they show lower hydrogen storage capacity than the other alloys. The advantages include easy usage and manufacturing, and high discharge-rate capability. Some AB_5 materials crystallise in the $CaCu_5$ structure and are most often modifications of the $LaNi_5$ structure, such as $LaNi_2Co_3$, $LaNi_2Co_{2.9}Al_{0.1}$, and $La_{0.5}Ce_{0.5}Ni_{3.55}Co_{0.75}Mn_{0.4}Al_{0.3}$. Changing the A and B composition can tailor the performance of the cell. The ratio between La and Ce is often changed depending on the cycle life and power capability required, and the amount of Co, Mn, and Al affects the activation and formation of the electrodes, resulting in increased performance of the final cell. Moreover, the production process of the active material also alters the final cell performance; the annealing process affects the capacity, discharge rate, and cycle life due to higher crystallinity obtained and higher purity at the grain boundaries. The capacity of the AB_5 alloys is about 300 mAh/g.

The AB_2 and A_2B_7 alloys are used in order to take advantage of higher capacity and higher energy density. The compositions of the AB_2 and A_2B_7 alloys are tailored for specific cell requirements and usage, usually based on: V, Ti, Zr, Ni, Cr, Mn, Sn, Co, Al, Mg, or La. The AB_2 alloys often have a capacity of about 450 mAh/g and the A_2B_7 alloys about 350–400 mAh/g.

For all metal alloys, the interface between the alloy and the electrolyte is important for the discharge capability and cycle life stability. The alloy used will gradually corrode in the strong alkaline electrolyte and will form oxide passivation layers on the surface. The corrosion also results in a less electronegative material for hydrogen storage and consumes water from the electrolyte. The latter leads to a gradual loss of power as the electrolyte is decomposed, thus increasing the cell resistance. Moreover, a gradual loss in capacity is observed due to the conversion of active material into corrosion products, which can migrate to the opposite electrode and activate poisoning and promote oxygen evolution. By optimising the alloy composition or by surface coatings, the corrosion rate can be controlled or hampered. Still, other factors affect the corrosion rate, including temperature, state of charge, and degree of overcharge.

The NiOOH positive electrode is fundamentally the same electrode as in the NiCd battery technology, and it can be made in different morphologies, e.g. spherical and sintered. As for most electrochemical cells, the performance is dependent on the active area of the NiOOH particles, due to the size of the electrode/electrolyte interface available for charge transfer. Moreover, the crystallinity, as well as the degree of impurities, affects the performance, why the manufacturing procedure is a key to the electrode design, and secure performance. NiOOH can exist in many forms, and β-NiOOH is the electrochemically preferred form. During discharge, β-Ni(OH)$_2$ is formed, which is reversibly transformed to β-NiOOH during charging.

The electrolyte is an aqueous solution of about 30% potassium hydroxide (KOH) (i.e. about 7M), providing high ionic conductivity. In order to enhance the charging efficiency by increasing the transference number (equation (1.23)), small amounts of, for example, LiOH and/or NaOH can be added to the electrolyte. This may also widen the operational temperature window even further towards higher temperatures. Through capillarity forces, the electrolyte is adsorbed in a separator, which is designed to keep as much electrolyte as possible without any overfill, and there is a limited amount of free electrolyte. The separator can, however, dry out because of slow phase transition and swelling of mainly the NiOOH material due to the incorporation of electrolyte in the electrode structure. This degradation process is temperature dependent and is further affected by cycling, as well as the type of electrode and electrode structure used. A dry-out of the electrolyte results in increased internal resistance, and subsequently a low capacity since the charge-transfer interface between the electrodes and electrolyte decreases.

NiMH cells may undergo more or less unexpected capacity losses, and the same phenomenon is observed in NiCd cells, often referred to as a *memory effect*. NiCd cells have been attributed reversible capacity losses if the cell is charged after being only partially discharged. The cell can 'remember' the capacity of the previous discharge, and will therefore only accept the same amount of capacity in the subsequent charging procedures. This can be crucial in the long-term if the cell is not fully discharged as large decreases in capacity may be observed. The cell can, however, recover if fully discharged. This effect is related to structural changes in the Cd electrode, and is therefore *not* an issue for NiMH cells.

On the other hand, the capacity losses observed in NiMH cells are related to *voltage depression*, which is not a memory effect, but rather due to long-term overcharge conditions, and may occur if the cell is repeatedly partially charged. A rapid decrease in voltage will be observed even if the capacity decrease is only slightly noticeable. The voltage depression is related to the NiOOH electrode, and is caused by a phase transition from β-NiOOH to γ-NiOOH, and is an issue for all Ni-based technologies. Upon discharge α-Ni(OH)$_2$ is reversible formed, and the subsequent cycling will utilise the γ-NiOOH/α-Ni(OH)$_2$ redox couple. α-Ni(OH)$_2$ can slowly transform back to the preferred β-Ni(OH)$_2$ phase at OCV conditions, as well as during a full discharge of the cell. As a consequence, NiMH cells should occasionally either be fully discharged or left at OCV.

2.3   Lithium batteries

Lithium (Li) is the lightest of all metals (0.54 kg/dm^3) and has a low electrochemical potential (−3.04 V vs. H$_2$/H$^+$), which makes it one of the most reactive metals. Together with a small radius (0.76 Å), which allows lithium to be accommodated in a variety of insertion host materials, these properties give lithium-based electrochemical cells the potential to achieve high-energy and power densities.

The lithium battery is a family of rechargeable batteries where lithium ions are the charge carriers that move back and forth between the electrodes during the charge/discharge processes. The cell potential is usually in the range of 2.5 V to 4.5 V.

Chemistry, performance, and safety characteristics vary widely within the family of lithium batteries, and can be optimised towards specific requirements. One of the family members, the lithium-ion battery, has become the most common rechargeable battery of choice for portable consumer electronics and electric vehicles having an all-electric driving range. In the following sections, the basics of lithium batteries will be described, and in Chapter 3 a more complete description of the lithium-ion technology is given.

Since there are several combinations of materials applicable for the electrodes, as well as for the electrolyte, the performance of lithium-based cells can be customised for specific applications and for capacity or rate, high-energy or power density. The cells in general have low weight and a high open circuit voltage, affecting energy density in a positive way. The cells can be cycled over a wide state of charge range, and some basic chemistries allow the cell voltage to be constant over a wide state of charge range. Fast charging is possible and the shelf life is acceptable.

For most lithium battery concepts, the temperature must be controlled, mainly due to the instabilities of the electrolyte, since side reactions may take place, and consequently cause capacity loss or thermal runaway (Section 6.1.2.1). An overcharging of the cell may cause the same phenomena. To reduce these risks, lithium batteries must contain a management unit, including a protective circuit, to control temperature, currents, and voltage levels in order to shut down the battery if an abuse situation arise. At low temperatures, in some cases already at $10°$ C, lithium batteries cannot be charged safely due to the increase of the internal resistance, and charging may cause side reactions and rapid degradation.

In the basic layout of the lithium battery, the two electrodes are materials allowing lithium ions to be inserted/extracted according to:

$$\text{Host} + \text{Li}^+ + e^- \Leftrightarrow \text{Li-Host}$$

Depending on the type of lithium battery, different host structures and materials are used; some combinations are more suitable for hybrid electric vehicles applications and others for electric vehicles. Due to the high cell voltage, water-based electrolytes cannot be used, and the electrolytes are instead based on a non-aqueous matrix: liquid, solid polymer, or a gel, in which a lithium salt is dissolved.

The most common lithium battery concepts are lithium metal, lithium-ion, and lithium-ion polymer. Lately, two other concepts based on lithium have had a revival: the lithium-oxygen and the lithium-sulphur batteries. In the following sub-sections, the basics for these concepts are described. Today, the Li-ion battery is the most widely used rechargeable battery. Therefore, a more detailed description of the Li-ion battery and its chemistry are to be found in Chapter 3.

2.3.1   Lithium metal

The first functional rechargeable lithium battery concept with high-energy density was the all-solid-state cell design using metallic lithium as the negative electrode, a lithium insertion material as the positive electrode, and a polymer-based electrolyte.

Due to the presence of lithium metal at the negative electrode, the cell can be assembled in the charged (Li-free positive electrode) or discharged (Li-containing positive electrode) state.

Metallic lithium has been disputed due to its involvement in abuse accidents. The main cause is arguably the growth of metallic lithium dendrites at the negative electrode during the charging process, which may reach across the electrolyte and finally cause an internal short circuit. Research has, however, been successful in finding ways to stabilise the metal surface and limit or inhibit dendrite formation. Lithium metal batteries with an operating temperature between 80° C and 120° C represent an attractive solution for electric vehicles.

2.3.2   Li-ion and Li-ion polymer

The Li-ion concept was designed to overcome the safety issues associated with the lithium metal cells. The negative electrode is commonly carbon based, typically graphite or hard carbon. Subsequently, the positive electrode must be lithium containing, usually an insertion material able to insert lithium in one, two, or three dimensions. Some widely used positive electrode materials are lithium-containing transition metal oxides, where the transition metal normally is Co, Ni, Fe, or Mn, or a combination thereof. The change in the metal's oxidation state makes the lithium insertion/extraction possible. The electrolyte used in Li-ion cells is typically a lithium salt solvated in an organic liquid solvent, all soaked in a separator for stability. When assembled, the Li-ion cell is in a discharged state, and lithium ions are extracted from the positive electrode and inserted into the negative electrode during the charge process, and vice versa during discharge.

The Li-ion polymer battery is very similar, and has most of the Li-ion concept benefits, but the electrolyte is a polymer matrix, gelled or plasticised by a liquid. The electrolyte is thus macroscopically perceived as a solid and the absence of free liquid makes these batteries more stable and less vulnerable to issues caused by overcharge, damage, or abuse. The electrolyte, also acting as a separator, cannot be made as thin as for the lithium-ion concept, thus limiting the effective surface area of the electrodes and hence the power capability of the cell.

2.3.3   Lithium-oxygen

The Li-oxygen (often and incorrectly named Li-air) concept is based on the oxidation of lithium at the negative electrode, and reduction of oxygen, ideally from the air, at the positive electrode. The major advantage of a Li-oxygen cell is the very high theoretical energy density possible, even though the practical energy density, all aspects considered, is perhaps more likely to be about two to three times higher than for Li-ion cells.

Primarily, four types of Li-oxygen batteries exist based on the electrolyte set-up: aprotic, aqueous, solid state, and mixed aqueous/aprotic – all use metallic lithium as the negative electrode. Although the electrochemical details vary depending on the choice of electrolyte, the practical cell potential is about 2.4 V. The positive electrode is often mesoporous carbon, incorporating a metal catalyst, as host material for oxygen. The capacity of the Li-oxygen cell is dependent both on the electrolyte and the catalyst. Catalysts based on Mn, Co, Ru, Pt, Ag, or a mixture of Co and Mn are being investigated.

A full reduction to Li_2O is preferred due to higher theoretical capacity (*ca.* 1800 mAh/g compared to *ca.* 1200 mAh/g for Li_2O_2). A major drawback is the insolubility of Li_2O_2 in aprotic electrolytes, which leads to clogging of the positive electrode. In an aprotic, non-aqueous electrolyte, the electrochemical reactions take place at the positive electrode according to one of the following reactions:

$$2Li + O_2 \rightarrow 2Li_2O$$

or

$$2Li + O_2 \rightarrow Li_2O_2$$

In an aqueous electrolyte, the reduction at the positive electrode depends on the pH, and in a basic environment the oxygen reduction reaction results in the formation of LiOH according to:

$$4Li + O_2 + 2H_2O \rightarrow 4LiOH$$

If the LiOH concentration is too high in the electrolyte, LiOH will precipitate as $LiOH \cdot H_2O$, and the corresponding theoretical capacity is *ca.* 640 mAh/g.

The Li-oxygen concept is mainly attractive for electric vehicles due to the high-energy density. The main drawbacks, though, are large voltage hysteresis, ΔE (as illustrated in Figure 2.4) and oxygen/air handling. The voltage hysteresis may be of the order of 1 V between charge and discharge, which would add constraints on the power electronics and other parts of an electric vehicle, and also result in a decrease in charge/discharge efficiency.

The oxygen/air handling may have even larger effects on the vehicle installation. If air is used, not only oxygen will approach the positive electrode; nitrogen, carbon dioxide, and water will also be present. These species will limit battery performance due to side reactions. Moreover, air pollutants may further inhibit the wanted electrochemical reactions. Therefore, clean and dry oxygen is preferred, preferably in the compressed state, adding further complexity to the battery and consequently reducing the energy density of the complete battery installation drastically. On account of these disadvantages, the Li-oxygen concept will not be discussed further.

2.3.4   Lithium-sulphur

Just like the Li-oxygen concept, the Li-sulphur concept utilises metallic lithium as the negative electrode, whereas the positive electrode employs elemental sulphur. The reduction of elemental sulphur (S_8) to a sulphide ion (S^{2-}) has a theoretical capacity of 1675 mAh/g, about ten times higher than for the lithium-ion. Practical values,

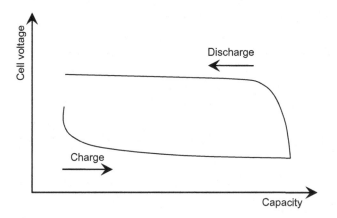

Figure 2.4 Voltage profile of a Li-oxygen cell, showing the hysteresis between charging and discharging.

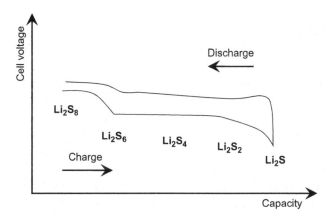

Figure 2.5 Schematic voltage profile for a Li-S cell.

however, stick between 500–800 mAh/g, depending on the discharge rates and whether or not all S is being utilised.

The cell voltage, though, is lower than for the lithium-ion; about 2 V. A schematic voltage profile is shown in Figure 2.5. Since sulphur itself lacks electronic conductivity, a carbon matrix is required to create a viable positive electrode, reducing the practical energy density.

During discharge, lithium from the negative electrode dissolves from the surface, diffuses through the electrolyte, and reacts with sulphur to form polysulphides at the positive (sulphur) electrode. In reverse, lithium plates on the negative electrode and polysulphides are reduced on the positive electrode surface sequentially during charging:

$$S_8 \rightarrow Li_2S_8 \rightarrow Li_2S_6 \rightarrow Li_2S_4 \rightarrow Li_2S_3 \rightarrow Li_2S_2 \rightarrow Li_2S$$

The high lithium-containing sulphides Li_2S_2 and Li_2S are insoluble in the electrolytes and do not take part in the redox reactions, resulting in a lower capacity than theoretically possible. While each sulphur atom can host up to two lithium ions, no insertion reactions take place at either of the electrodes, in contrast to other types of lithium batteries.

One of the main drawbacks of Li-S batteries is the electrolyte being both unstable and solubilising polysulphides. While sulphur and Li_2S are relatively insoluble in most electrolytes, many of the intermediary polysulphides hence cause irreversible loss of active material.

2.4   Electrochemical double-layer capacitors

Electrochemical double-layer capacitors (EDLC), also known as *super capacitors* or *ultra capacitors*, are high-energy density cells having a lot in common with battery cells. The basic difference is that a capacitor is an electrical energy storage device while a battery is a chemical energy storage device. Up till now most applications have been equipped with batteries due to their large energy storage capabilities. For many vehicle applications, though, power, not only energy, is of key importance in terms of performance and driveability, exceeding the capabilities of battery technologies. In order to be attractive as a capacitor for vehicle applications, the capacitor must possess high-energy density, be small in size, and have a long shelf and cycle life. As for batteries, there is a trade-off between power and energy density, as shown in the Ragone plot in Figure 1.28. The discharge time is much shorter for a capacitor than for a battery.

In a capacitor, energy is stored by a charge separation in an electrostatic field between the electrodes. The ions in the electrolyte migrate towards the electrodes of opposite polarity, and charge separation takes place in the double-layer formed at the interfaces between the liquid electrolyte and the micro-porous surfaces of the electrodes. Since no chemical reactions are involved, very fast charge/discharge processes are possible.

A capacitor is hence very similar to a battery cell: two electrodes, a separator, and an electrolyte. The difference is that micro-porous materials are commonly used for the electrodes to maximise the surface area. The separator needs to be permeable for the ions in the electrolyte, the electrolyte usually being a liquid, either organic or aqueous, or an ionic liquid. Figure 2.6 illustrates a schematic capacitor.

The cell voltage for a capacitor based on an aqueous electrolyte is about 1 V and when based on an organic electrolyte about 3 V. Ideally, the cell voltage is directly proportional to the state of charge, as schematically shown in Figure 2.7. At the beginning of discharge, the voltage drops quickly as a consequence of the internal resistance of the cell – the *IR drop*. Due to the linear discharge voltage characteristics and the limited available energy density, appreciable power can only be delivered for very short durations, normally a few tens of seconds.

The energy contained by charge separation takes place in a double-layer with a small thickness, resulting in a high capacitance. The amount of energy that can be stored is given by:

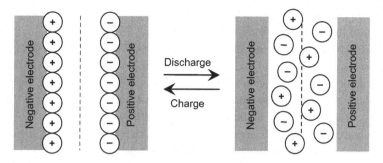

Figure 2.6 Illustration of an EDLC.

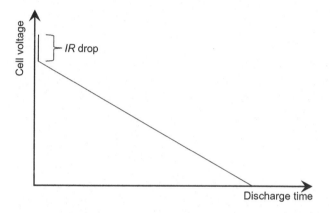

Figure 2.7 Schematic discharge characteristics of an EDLC.

$$Energy = \frac{1}{2}CV^2 \tag{2.2}$$

where V is the cell voltage and C is the *capacitance* (measured in *farad*, F). In an ideal double-layer capacitor, the capacitance is calculated based on the geometry of the electrodes and the dielectric properties of the electrolyte according to the following equation:

$$C_{dl} = \varepsilon \frac{A}{d} \tag{2.3}$$

where A is the active electrode area, ε is a constant (the dielectric constant or the permittivity of the electrolyte), and d is the charge separation distance. The pore size of the electrodes thus plays an essential role and should be large enough to incorporate the ions in order to use the entire available electrode surface. Ideally, the capacitance is constant and independent of the cell voltage and no faradic reactions take place between the electrodes and the electrolyte. This *double-layer capacitance*, C_{dl}, is electrostatic storage achieved by charge separation in a Helmholtz double-layer (Section 1.5.2) with a separation distance of a few Ångstroms.

The capacitance is often difficult to accurately determine due to the active area being a function of the pore sizes. The maximum voltage of a capacitor depends on the breakdown characteristics of the dielectric material, in this case the electrolyte.

The stored energy is therefore a function of size, distance, and material properties of the surfaces of the electrodes, and the cell voltage.

The double-layer is formed and relaxed almost instantaneously, having a formation time constant of the order of 10^{-8} seconds. Therefore, the structure of the double-layer has the ability to respond rapidly to changes in the cell compared to an electrochemical cell, where the redox reactions time constants are somewhere between $10^{-2} - 10^{-4}$ seconds, which are related to the impedance of the reactions.

Utilising capacitors as energy storage for electric vehicles means there are some fundamental differences compared to battery cells when connecting the capacitors in series and/or parallel configurations. When adding n cells in a parallel configuration, the total active electrode surface area is increased by the number of cells, resulting in a sum of capacitance according to:

$$C_{tot} = C_1 + C_2 + \ldots + C_n \tag{2.4}$$

On the other hand, connecting n cells in series, the separation distance, d, increases, not the electrode surface area. The total voltage difference is proportional to each capacitor according to the inverse of its capacitance. The entire series acts as a capacitor with a lower capacitance than any of its components:

$$\frac{1}{C_{tot}} = \frac{1}{C_1} + \frac{1}{C_2} + \ldots + \frac{1}{C_n} \tag{2.5}$$

Unlike batteries, the voltage drops significantly during discharge (Figure 2.7). Therefore, energy recovery requires complex electronic control and balancing equipment. The capacitors can use the full voltage range for power delivery all the way down to 0 V without any damage to the cell, and therefore it is common to transport and store the capacitors fully discharged and in a short circuit state.

For all-electric vehicle applications, both energy and power are required, although in different ratios. By combining the benefits of a capacitor with a battery, another type of energy storage is enabled. In such a combined system, battery life can be significantly extended through optimal usage of the two components. This combination will, however, most likely require a DC/DC converter in order to be functional.

The capacitance can also be of pseudo-capacitance character; interactions between the electrodes and the electrolyte involving faradic reactions of charge-transfer character. The energy is achieved by redox reactions, electrosorbtion on the surface of the electrode by specifically absorbed ions, resulting in a reversible faradic charge transfer of the electrode. The charge transfer is dependent on the voltage and consequently the pseudo-capacitance C_{ps}:

$$C_{ps} = \frac{dQ}{dV} \tag{2.6}$$

Depending on the materials chosen for electrodes and electrolyte, different kinds of high-energy density capacitors can be tailored to suit a variety of applications and needs. The ratio of the two types of capacitances, C_{dl} and C_{ps}, can vary greatly, and three main categories exist:

- *Double-layer capacitors* – with carbon electrodes or derivates with high double-layer capacitance: the 'traditional' double-layer capacitors (EDLCs).
- *Pseudo-capacitors* – with electrodes made of metal oxides or conducting polymers with high faradic pseudo-capacitance.
- *Asymmetric capacitors* – with one electrode of double-layer capacitor character and one of pseudo-capacitor character.

In the case of asymmetric capacitors, often called hybrid capacitors, the energy storage capacity is normally higher than for the double-layer capacitor. The asymmetric capacitor can also be designed using two different pseudo-capacitor style electrodes.

2.4.1   Capacitor materials

When designing capacitors of high capacitance, electrode materials with very large surface areas are required and this is normally achieved using thin nano-sized carbon coatings on a metallic current collector. The contact resistance between the carbon material and the current collector should be as low as possible to reduce the overall cell resistance and increase the power capability. Activated carbon is the dominating coating material, but more and more graphitised carbon, nano-tubes, or graphene are used to obtain increased capacitance.

For the pseudo-capacitors, metal oxides are used as electrode materials, normally RuO_2 or MnO_2, and the redox reactions take place at the metal-oxide surfaces of both electrodes. The metal oxide is mixed with carbon in order to increase the electric conductivity (similar to the composite electrodes of battery cells). For composite electrodes, it is important to distinguish between the capacitance of the active material and that of the final electrode, which is normally lower.

The maximum cell voltage possible, for any capacitor design, depends on cell resistance and type of electrolyte: commonly aqueous or organic solvent based. An aqueous electrolyte, e.g. H_2SO_4 or KOH solutions, gives a cell voltage of about 1 V. Organic electrolytes, mainly acetonitrile or propylene carbonate based with a salt added for ionic conductivity, result in a cell voltage of about 2.5–3 V.

2.4.2   High-energy capacitors

There are two fundamental ways to increase the energy density of a capacitor: by increasing either cell voltage or capacitance (equation (2.2)). One way to increase the cell voltage is by changing the type of electrolyte. Another way is to utilise the advantages of asymmetric capacitors by employing both faradic and non-faradic processes. To couple a redox material with a high capacitance material results in higher operational cell voltage and higher cell capacitance. An attractive approach is to use activated carbon as one electrode and an insertion electrode of a Li-ion cell as the other, so-called Li-ion capacitors. Utilising a graphite insertion electrode will result in an increased operational voltage range, as illustrated in Figure 2.8. For comparison, a 'symmetric' EDLC using activated carbon is also shown. The high operational voltage

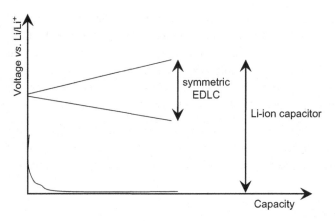

Figure 2.8 Operational voltage range for (a) a 'symmetric' activated carbon EDLC and (b) a Li-ion capacitor utilising activated carbon and graphite.

enables Li-ion capacitors of high-power and high-energy density; they are one type of pseudo-capacitors. Due to the instability of graphite in organic electrolytes (Section 3.1.3), other types of lithium ion insertion electrode materials have been considered, e.g. nano-sized $Li_4Ti_5O_{12}$ (Section 3.1.5.1), because of their flat voltage profile and their stability towards the electrolyte.

Ionic liquids have been shown to enable a cell voltage of about 4 V, and consequently the energy density will increase by about 80% compared to a 3 V cell if the same capacitance is achieved (since the energy density is proportional to V^2 (equation (2.2))). The power capability of the cell depends on the ionic conductivity of the electrolyte, which affects the cell resistance. Ionic liquids often have low conductivities at low temperatures, which is why elevated temperatures are often needed.

2.5   Other battery technologies

The above described battery and capacitor technologies are the most common and the most suitable technologies for electric vehicles. This is primarily due to their basic properties, but also to the well-established and robust production from different manufacturers. There are, however, other battery technologies, that could be viable, depending on application and market perspective.

All-electric vehicles based on fuel cells (Section 2.6) also represent promising progress. A fuel cell is in basic terms not a battery since no energy is stored in the cell; it is an electrochemical energy converter, and can also be compared to redox flow batteries.

2.5.1   High-temperature molten-salt batteries

Redox couples operating at high temperatures (i.e. above 100 °C) create an opening for other types of battery technologies. One group of rechargeable high-temperature cells are based on molten salts as the electrolyte. This enables high-power densities due to fast kinetics, and by carefully selecting the active materials high-energy density can also be achieved. The operating temperatures of these battery technologies are in the range of 200–700 °C, requiring different thermal and safety management than technologies operating at room temperatures.

The most common technology is based on sodium: sodium/sulphur and sodium/nickel cells. The latter, commonly known as the *ZEBRA* battery, will be used to illustrate the high-temperature molten-salt technology. The name refers to the research project Zeolite Battery Research Africa starting in the mid 1980s, and is the name used in the battery community. The technical name is, however, the Na-NiCl$_2$ battery, and a schematic cell is illustrated in Figure 2.9.

In a Na-NiCl$_2$ cell, molten sodium (melting point 98 °C) is used as the negative electrode, and the positive electrode consists of nickel in the discharged state (i.e. the pure metal) and NiCl$_2$ in the charged state. The electrolyte comprises molten sodium aluminium chloride (NaAlCl$_4$), which has a melting point of 157 °C. Below the melting point of the NaAlCl$_4$ electrolyte, the cell is unusable. The operating temperature is within the range of 220–450 °C, normally in the range of 270–350 °C. Nickel and NiCl$_2$ are almost insoluble in neutral and basic melts, and contact between these compounds is therefore possible, resulting in low charge-transfer resistance. NaAlCl$_4$ and Na are liquid at the operating temperature, and a separator of a Na-conducting β-Al$_2$O$_3$ ceramic is used to separate the liquid sodium from the molten NaAlCl$_4$.

During the discharge process, NiCl$_2$ is reduced to Ni, and consequently Na is oxidised to NaCl:

$$NiCl_2 + 2Na \rightarrow Ni + 2NaCl$$

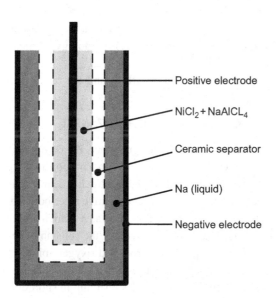

Positive electrode

NiCl$_2$ + NaAlCL$_4$

Ceramic separator

Na (liquid)

Negative electrode

Figure 2.9 Schematic illustration of a Na-NiCl$_2$ cell.

The reverse reaction takes place during the charging process. The nominal cell voltage is about 2.6 V, but at the end of discharge, when all $NiCl_2$ is consumed, the voltage drops quickly. However, the discharge can proceed and eventually results in an over-discharge reaction; the reduction of $NaAlCl_4$ to aluminium:

$$3Na + NaAlCl_4 \rightarrow 4\,NaCl + Al$$

This reaction occurs at an OCV of about 1.6 V and is reversible upon recharging of the cell. The associated voltage drop can be used for monitoring the end of normal discharge to avoid overdischarge. $NaAlCl_4$ is also involved in the reaction occurring when an overcharged state has been reached; $NiCl_2$ will continue to be formed even after all NaCl has been consumed, according to the following reaction:

$$Ni + 2NaAlCl_4 \rightarrow 2Na + 2AlCl_3 + NiCl_2$$

Overcharge takes place at an OCV of about 3.0 V, and is, like the overdischarge reaction, reversible. During excessive overcharge, the positive electrode will start to degrade.

The ZEBRA technology is a potential battery technology for vehicle applications, but due to the high operating temperatures, safety is one of the concerns. Normally, the cells are encapsulated in a thermos-like compartment and the temperature is conserved at a high level over a long period of time. The vibrations arising during vehicle usage may induce defects in the sensitive β-Al_2O_3 ceramic separator.

2.5.2   Nickel zinc batteries

The NiZn technology has much in common with the NiMH technology. They both utilise NiOOH as the positive electrode and an aqueous alkaline electrolyte. Compared to NiMH, the NiZn battery has an OCV of about 1.7 V (compared to 1.2 V for NiMH), resulting in enhanced energy density. A Zn electrode having a large surface area enables fast electrochemical kinetics and, subsequently, the NiZn technology is suitable for high rate demanding applications, such as hybrid electric vehicles.

During discharge, Zn is ideally oxidised to $Zn(OH)_2$, and the overall cell reaction during discharge is:

$$2NiOOH + Zn + H_2O \rightarrow 2Ni(OH)_2 + Zn(OH)_2$$

Depending on discharge conditions and cell design, other Zn compounds may be formed, e.g. $ZnO_2{}^{2-}$ or ZnO, which are converted back to Zn upon recharge. As can be noted, the water in the electrolyte takes part in the electrochemical reaction.

If the cell is overcharged, hydrogen is formed at the negative electrode and oxygen at the positive electrode, due to the splitting reaction of water, which might occur at different reaction rates. On the other hand, at overdischarge the cell can be reversed and the same reactions may occur, but at opposite electrodes. If oxygen migrates to the Zn electrode, a rapid reaction between oxygen and Zn to form ZnO will take place. During recharge, zinc may form dendrites across the electrodes resulting in short circuiting and possible abuse conditions.

The self-discharge rate for the NiZn technology is enhanced due to the solubility of zinc in the alkaline electrolyte. Zn is thermodynamically unstable and the following reaction takes place:

$$Zn + H_2O \rightarrow ZnO + H_2$$

It is possible to improve the stability of the Zn electrode by introducing other elements or through adding inert materials or additives, or altering the concentration of the electrolyte. It is important to match these performance enhancers, since some of them improve the Zn electrode performance while, at the same time, impairing the NiOOH electrode performance. For example, the capacity of the NiOOH electrode is maximised by a high OH^- concentration in the electrolyte, which significantly increases the solubility of Zn.

2.5.3   Zinc-air batteries

Zinc-air batteries are usually related to small primary cells, like button cells for hearing aids. Rechargeable Zn-air cells, by mechanically changing the electrodes, have been available for some time. Electrically rechargeable Zn-air cells have been developed more recently and are a potential technology for electric vehicles. Such a Zn-air cell comprises a three-phase concept: solid zinc metal as the negative electrode, a bifunctional air electrode as the positive electrode, and a liquid alkaline-based electrolyte in a separator. The electrolyte is usually of the same type as for nickel-based technologies: a high concentrated aqueous solution of potassium hydroxide.

The air electrode is divided into a catalytic active layer and a porous gas diffusion layer, which is permeable to oxygen and is hydrophobic. The oxygen pressure difference between the atmosphere and the inside of the cell is utilised for operation. The catalyst facilitates the reduction of oxygen to hydroxyl ions, which migrate from the positive to the negative electrode to complete the redox reaction. The chemical reactions during discharge are the following reactions:

$$\text{Negative electrode: } Zn + 4OH^- \rightarrow Zn(OH)_4^{2-} + 2e^-$$

$$Zn(OH)_4^{2-} \rightarrow ZnO + H_2O + 2OH^-$$

$$\text{Positive electrode: } O_2 + 2H_2O + 4e^- \rightarrow 4OH^-$$

$$\text{Overall reaction: } 2Zn + O_2 \rightarrow 2ZnO$$

The cell potential is theoretically 1.65 V, but in order to obtain long-lived cells the actual potential is kept at 1.2–1.4 V. The shape of the voltage profile has much in common with the voltage profile of NiMH technology.

Recharging, i.e. to liberate oxygen from ZnO, is difficult, while the charging potential is higher than the discharge potential. This may result in an energy efficiency cycle as low as 50%. It is also crucial to understand the interactions between the $Zn(OH)_4^{2-}$ ions and the alkaline electrolyte to limit capacity losses. One of the side reactions is the hydrogen evolution reaction:

$$Zn + 2H_2O \rightarrow Zn(OH)_2 + H_2$$

This reaction may occur at the zinc electrode in the alkaline electrolyte, but fortunately there are design options to overcome this issue by coating, alloying, or adding additives to the electrolyte. From a cell design perspective, a critical issue is the design of the positive electrode, including the catalysts. The catalysts are essential in order to decrease the overpotential of the electrode (and improve cycle energy efficiency). During operation, the oxygen flow into the cell has to be controlled in order to avoid drying out of the cell. Moreover, migration of Zn^{2+} from the negative to the positive electrode may occur, causing a capacity decrease. Therefore, the pore size of the separator should be optimized. Another issue of air electrodes is CO_2 interacting with the alkaline electrolytes to form carbonates, which reduce the conductivity of the electrolyte.

2.5.4   Metal-ion batteries

Based on the same basic principles as the Li-ion concept, other types of metal-ion (Me-ion) concepts are possible, e.g. Na-ion, Mg-ion, and Al-ion (the two latter more often and correctly named Mg-metal and Al-metal). The Me-ion concepts of relevance depend on the electrochemical capacity and the operating voltages. The theoretical capacity is related to the number of electrons involved in the redox reactions (Section 1.6.3, equation (1.28)), and therefore it is of interest to use multivalent ions to double or even triple capacity if possible. Thus, despite their larger atomic weights, magnesium- and aluminium-based concepts are attractive because of their ability to exchange two and three electrons, respectively, per ion compared to only one electron for lithium and sodium. The practical capacity in turn depends on the amount of reversible ions inserted/extracted during the charge and discharge processes. When using multivalent metal ions, e.g. Mg^{2+} or Al^{3+}, only half or one third of the ions are required to enable the same capacity as for a monovalent metal ion.

How suitable a technology or concept is for vehicle application is related to the electrodes' ability to reversibly cycle Me-ions at acceptable rates and to have a capacity comparable to other battery technologies. In the reversible reactions, a large amount of cyclable ions at suitable voltage levels should be used to give high-energy density and exhibit high ion-diffusion rates in order to meet the power requirements of different driving conditions. Some basic parameters are listed for the different metals of interest in Table 2.1.

The ion radius affects the insertion/extraction performance of the material used. Even if some metal ions have similar radii, the insertion processes may differ, and the host structure suitable for one type of Me-ions may not be suitable for another. The interactions, mainly of Coulombic character, between the Me-ion and the host structure determine cell performance and the ion diffusion properties.

Na-ion is one of the more attractive candidates and the concept is comparable to the Li-ion concept in many aspects; the voltage levels are in the same range and the energy density is comparable with Li-ion batteries and thus of interest for electric

Table 2.1 Summary of metals relevant for Me-ion cells with respect to basic properties (lithium is included for comparison)

Metal	Atomic weight (g/mole)	Redox couple	Standard reduction potential (V)	Ion radius (Å)	Theoretical capacity (mAh/g)
Lithium	6.94	Li^+/Li	−3.0	0.76	3860
Sodium	22.99	Na^+/Na	−2.7	1.02	1170
Magnesium	24.31	Mg^{2+}/Mg	−2.4	0.72	1340
Aluminium	26.98	Al^{3+}/Al	−1.7	0.54	2980

vehicle applications. Moreover, the availability of sodium in the Earth's crust is more than a 1000 times higher than that of lithium, resulting in a more solid sustainability perspective and long-term cost competitiveness for the Na-ion concept. A drawback of Li-ion cells is the alloying of lithium with aluminium at low potentials, and the reason why copper must be used as the negative electrode current collector (Section 4.1). Sodium, on the other hand, does not form alloys with aluminium; hence aluminium can be used as the current collector for both electrodes, resulting in a lower total weight of the Na-ion cell compared to the Li-ion cell. Due to the larger radius of Na^+ compared to Li^+, however, larger interstitial sites in the host structure are required for facile insertion.

Electrode materials of interest are primarily those having low activation polarisation for Na^+ transport and small volume changes during insertion and extraction. The main groups of material for the positive electrode are: layered transition metal oxides and polyanionic framework structures. The layered materials are often of an Na_xMO_2 ($x \leq 1$) insertion character, where M is a transition metal, e.g. Mn, Co, Fe, or Cr, or a mixture thereof. The polyanionic framework structures often contain PO_4 groups, either alone or in combination with F. Examples are $Na_3Ti(PO_4)_3$, $NaVPO_4F$, and $Na_2(Fe/Mn)PO_4F$. In some of these cases, two Na^+ ions can be reversibly cycled, thereby increasing the theoretical capacity of the cell by a factor two.

As for the negative electrode, mainly carbon-based materials are of interest, but alloys or conversion reaction materials may be used. The most commonly employed active material for the negative electrode of Li-ion cells, graphite, is not able to incorporate Na^+ ions into the structure, which is surprising since both Li^+ and K^+ can be inserted into graphite. Therefore, disordered carbon materials are used for Na-ion negative electrodes with diverse morphologies, microstructures, and degrees of graphitisation. One example is hard carbon, a material where the Na^+ ions are inserted into nano-pores between randomly stacked graphene layers. Other types of negative electrode materials are alloys, mainly Sn, Sb, or Sn/C. The large ionic radius of the Na^+ ion is, however, expected to result in large volume changes upon formation of sodium alloys.

Considering the electrolyte, Na-ion cells have the same demands as every other battery technology e.g. it must be stable in the whole voltage range in order to secure durability and safe usage. The electrolytes used in Na-ion cells are very similar to the ones used in Li-ion cells, which are further discussed in Section 3.3.

Moving to multivalent alternatives, utilising the alkaline earth metal magnesium would be a step towards two-electron redox reactions that would 'easily' double capacity compared to the Li-ion or Na-ion cells. Magnesium is an abundant metal in the Earth's crust and has a low reduction potential (Table 2.1). The Mg concept uses metallic magnesium as the negative electrode, which does not suffer from dendrite formation to the same extent as metallic lithium. Just like metallic lithium, however, magnesium forms a passivation layer in organic non-aqueous electrolytes that severely reduces the Mg^{2+} ion diffusion rates to the electrode. Many types of electrolytes, liquid, polymer gel, and ionic liquids can be applied, but the corrosive nature of the electrolyte must be considered in the selection. Examples of electrolytes of interest are organohaloaluminiates and organoborates of Mg.

The main challenge of Mg-based batteries, and of all multivalent concepts, is not, however, on the negative electrode side but on the positive. This electrode requires materials that allow several oxidation state steps and acceptable diffusion rates of the high charge density Mg^{2+} cation. The ideal material would be a material based on the transition metals having one reversible redox couple per inserted Mg^{2+}, i.e. a two-electron reduction of the transition metal, e.g. vanadium, manganese, or titanium. The most promising materials are insertion compounds based on oxides or sulphides, due to their capacity and potential.

The most studied group of materials for the positive electrode is the Mg-based Chevrel phases $Mg_xMo_6T_8$, where T is S, Se, or a mixture thereof. The structure consists of octahedrally coordinated Mo in a cubic framework of the anions S and/or Se. The Mo atoms exhibit variable oxidation states and the anion framework provides diffusion pathways in several directions and a variety of sites for the inserted Mg^{2+} ions. Up to four electrons can be sustained by the Mo_6 clusters, resulting in a theoretical capacity of up to two Mg^{2+} ions per Mo_6T_8 unit. Factors affecting the ability of the material to incorporate Mg^{2+} are primarily the solid-state diffusion rates, which, if not favourable enough, will increase the electrode polarisation and may cause ion-trapping. Secondary factors are possible co-insertion of the electrolyte solvent and concomitant structural distortion or decomposition of the positive electrode material. Yet, many such materials suffer from low electronic conductivity and blended materials may therefore be used. The Chevrel phase materials enable rapid Mg^{2+} diffusion rates due to the large amount of vacant sites available in the structure, and the diffusion rates can be enhanced at elevated temperatures.

2.5.5   Redox flow batteries

Redox flow (or flowing electrolyte) batteries have been developed since the late 1960s. The technology represents a more complex battery design. The key to the technology is that the redox reactions involve only ions in solution and the reactions take place at inert electrodes. The active materials are thus contained in the electrolyte and stored externally and pumped past the electrode. Generally, the redox flow technology possesses long durability, but energy density is low due to limited solubility of the active materials and several auxiliary components, e.g. pumps and storage tanks. The vanadium redox

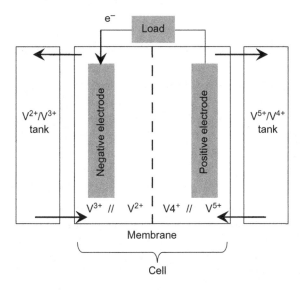

Figure 2.10 A simplified vanadium-redox flow system.

flow battery is one of the most common and most studied, and will serve as an example to demonstrate the technology.

To store energy, the concept employs vanadium ions in different oxidation states: V^{5+}/V^{4+} at the positive side and V^{2+}/V^{3+} at the negative side. Figure 2.10 shows the principle of a vanadium-redox flow system in a simplified and schematic manner.

The vanadium redox technology is based on the ability of vanadium to exist in four different oxidation states: II, III, IV, and V. The electrolytes are commonly aqueous-based sulphuric acid solutions. A typical OCV is about 1.4 V at 25° C, and the negative and positive electrode reactions, respectively, during discharge are:

$$\text{Negative electrode: } V^{2+} \rightarrow V^{3+} + e^-$$

$$\text{Positive electrode: } V^{5+} + e^- \rightarrow V^{4+}$$

The *two* electrolytes are pumped through the cell to facilitate the electrochemical reactions and an ion-selective membrane is used to separate the two electrolyte compartments. Mixing the electrolytes will result in a permanent loss of energy storage capacity due to dilution of the active materials.

One of the main advantages of redox flow technology is the almost unlimited energy storage capacity; the size of the storage tanks defines the limits. Moreover, it can be left in a completely discharged state without risking damage, and it can be recharged by replacing the vanadium solutions or by applying an external power source. The low energy density is the main disadvantage of the technology along with system complexity. Another limitation is the temperature range of operation: 10–40° C. At lower temperatures, the ion-infused sulphuric acid crystallises. However, using vanadium/bromine redox flow batteries may increase the energy density and widen the temperature range. In a battery, cells are assembled in a bipolar manner (Section 1.3.2) to increase the voltage.

2.6   Fuel cells

A fuel cell is a galvanic electrochemical cell converting the chemical energy of a fuel into electricity when reacting with an oxidant. The fuel cell cannot store energy; energy can only be converted and the essential difference between a fuel cell and a battery cell is the supply of energy source. Fuel and oxidant are continuously supplied from external sources when required, and thus the technology has much in common with the redox flow battery technology.

Just like a battery cell, the fuel cell consists of a positive and a negative electrode separated by an electrolyte, but the electrodes of a fuel cell are inert. Instead, the electrodes have catalytic properties enhancing the reactions. The active material of the negative electrode is the fuel, fed to the electrode in gaseous or liquid form and the most commonly used fuels are hydrogen, natural gas, methanol, or other types of hydrocarbons. Similarly, an oxidant is fed to the positive electrode and the most commonly used oxidant is oxygen, preferably from the air. A schematic fuel cell is shown in Figure 2.11.

The electrodes are separated by the electrolyte, in fuel cells often called the membrane. Catalysts are added to the electrodes to facilitate the electrochemical reactions. As fuel (hydrogen) and oxidant (oxygen) are fed, and the cell is connected to an external load, the following reactions take place:

$$\text{Negative electrode: } H_2 \rightarrow H^+ + e^-$$

$$\text{Positive electrode: } O_2 + 4H^+ + 4e^- \rightarrow 2H_2O$$

$$\text{Overall cell reaction: } 2H_2 + O_2 \rightarrow H_2O$$

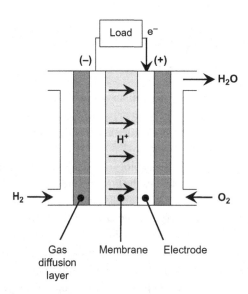

Figure 2.11 Schematic illustration of a fuel cell.

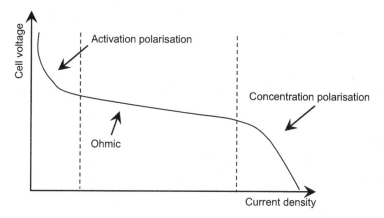

Figure 2.12 Polarisation profile for a fuel cell.

The performance of a fuel cell is often visualised by plotting the cell voltage versus the current, or current density; a *polarisation profile*, as illustrated in Figure 2.12. The shape of the profile is affected by the concentration polarisation, η_c, the activation polarisation, η_a, of the electrodes, the internal resistance or ohmic resistance, R, and the current density, i, according to the same relationship as for a battery cell (Section 1.6.1). Ideally, in the absence of any limiting parameters, the theoretical voltage of an H_2/O_2 fuel cell would be 1.23 V, i.e. the disproportional potential of water. During real fuel cell operation, the voltage is lower though.

As indicated in Figure 2.12, at low current densities the activation polarisation limits cell performance and at high current densities the concentration polarisation is the limiting factor. In between, at moderate current densities, the internal ohmic resistance limits the performance.

Different types of fuel cell concepts exist, and their classification is based on the fuel, the electrolyte, and the operational temperature. The most common types are summarised in Table 2.2. The PEM fuel cell will be used to illustrate the benefits and limitations of fuel cell technology.

2.6.1   Polymer electrolyte membrane fuel cells

The *polymer electrolyte membrane fuel cell* (PEMFC) utilises hydrogen as fuel and oxygen as oxidant, according to the working principles described above. When dissociated at the negative electrode, the protons are transported through the membrane and recombined with electrons and oxygen at the positive electrode to form water, the only waste produced by a PEMFC.

The power density of any fuel cell is related to the active electrode area, and the PEMFC shows a high-power density (*ca.* 1 W/cm^2), which is attractive for vehicle applications. The operational temperature is low, for a fuel cell, normally in the interval of 80–120 °C. On the other hand, the performance is limited by the purity of the fuel and the catalysts. Each cell will have an effective operational voltage of about 0.7 V,

Table 2.2 The most common fuel cell concepts and their key characteristics, advantages and disadvantages.

	Polymer electrolyte membrane (PEMFC)	Direct methanol (DMFC)	Phosphoric acid (PAFC)	Alkaline (AFC)	Molten carbonate (MCFC)	Solid-oxide (SOFC)
Electrolyte	Fluorinated organic polymer	Fluorinated organic polymer	Phosphoric acid	Alkali hydroxide (KOH, aq)	Molten carbonate	Yttria-Zirconia
Temp. range (° C)	70–200	60–90	Ca. 200	25–220	Ca. 550	600–1000
Fuel	H_2	Methanol	H_2	H_2	H_2 or CH_4	H_2 or CH_4 or other hydrocarbons
Charge carrier	H^+	H^+	H^+	OH^-	CO_3^{2-}	O^{2-}
Advantages	High-power density, fast start-up, suitable for vehicle applications	Simple fuel storage	High fuel efficiency, low sensitivity to impurities, suitable for stationary applications	High fuel efficiency	High fuel efficiency, low sensitivity to impurities, suitable for continuous power needs	High fuel efficiency, low sensitivity to impurities
Disadvantages	Sensitive to CO, fuel storage	Toxic fuel, low power density	Slow start-up, low power density	Sensitive to impurities (e.g. CO_2)	Poor durability, high-temperature degradation	Slow start-up, high-temperature degradation

and in order to reach the desired voltage, several cells are stacked in series forming a fuel cell stack of bipolar design (Section 1.3.2). The cell itself consists of four main components: the *bipolar plate*, the *gas diffusion layer*, the *electrodes*, and the *polymer electrolyte membrane*, as indicated in Figure 2.11.

2.6.1.1   Bipolar plates and gas diffusion layers

The bipolar plate has several functions within the fuel cell. First of all it distributes fuel and oxidant gases over the entire active area, and it conducts the electrons via the external circuit to the load. It also functions as mechanical support to the whole stack. Since the stack is assembled in a bipolar manner, the bipolar plate acts as negative electrode in one cell and as positive in the other. Moreover, the bipolar plate will facilitate the heat transfer from the cell and if needed it often incorporates the cooling medium.

The bipolar plates are made of materials enabling high electronic conductivity: composites or metals. The composite has the advantage of being chemically inert in the strong acidic environment, especially at the negative electrode. On the other hand, composites are often brittle, requiring thicker bipolar plates, which result in larger and heavier fuel cell stacks. Metallic plates can be manufactured as thinner plates, but are not chemically tolerant and often corrode in the fuel cell environment. Depending on the metal or alloy used, the electric conductivity may be lower than for the composite plates. Corrosion of the bipolar plates not only affects stability and gas flow, it will also limit the performance of the catalyst, as the corrosion products will act as poisoning agents.

As the bipolar plates are responsible for the gas flow to and along the active electrode area, it is important to design plates facilitating a gas flow with as small a pressure drop as possible. Therefore, the plates are equipped with different flow channel patterns to cover the whole active area with gas. Moreover, for the positive electrode the gas flow channels should also be able to remove the water produced. In order to increase gas distribution, a *gas diffusion layer* is incorporated between the bipolar plate and the electrode. The layer is usually a carbon-based cloth or fibre network designed for gas permeability and high electronic conductivity and its main task is to distribute the gas evenly over the entire active area and to transport the water produced away from the surface and further to the outlet. Therefore, the hydrophobic character of the layer is essential.

2.6.1.2 Electrodes

In contrast to a battery cell, the electrodes in a fuel cell are inert and do not take part in the electrochemical reactions. The electrodes are composites of catalyst particles, carbon support, and a proton-conducting matrix. The catalyst enables the redox reactions to proceed at a high rate and at a moderate temperature. Carbon support has the same function as in a composite electrode for battery cells: to facilitate the electronic conduction between the catalyst particles and the gas diffusion layer and the bipolar plates. The glue to make the electrode manageable is the proton-conduction matrix, which is also responsible for proton transport between the different reaction sites and the polymer electrolyte membrane (PEM) (see below).

The catalyst is usually based on platinum (Pt); the only known catalyst capable of splitting hydrogen and the recombination with oxygen to water at high rates. Therefore, Pt-based catalysts are used at negative and positive electrodes. Improved cell performance is obtained if the electrodes are deposited directly on to the PEM, or the gas diffusion layer, in order to receive electrodes as thin as possible. Often the gas diffusion layers, the electrodes, and the PEM are manufactured as one single unit that is easier to handle: the *membrane electrode assembly* (MEA).

2.6.1.3 Polymer electrode membrane

The PEM is the equivalent of an electrolyte in a battery cell whose functions are to transport protons from the negative to the positive electrode and to block any mass transport in the other direction. Moreover, the PEM should be an electric insulator, like all electrolytes. The material used is commonly a polymer of a poly-fluorosulfonic acid, with side chains of different lengths, as shown in Figure 2.13. Examples of a long-chain polymer is Nafion®. The backbone of the polymer has a hydrophobic character and the end-groups of the side-chain are of hydrophilic character, absorbing water to

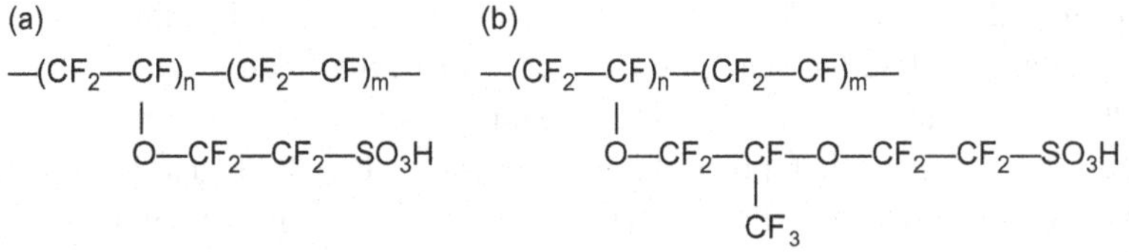

(a)   (b)

—(CF$_2$—CF)$_n$—(CF$_2$—CF)$_m$—   —(CF$_2$—CF)$_n$—(CF$_2$—CF)$_m$—

     | |

    O—CF$_2$—CF$_2$—SO$_3$H   O—CF$_2$—CF—O—CF$_2$—CF$_2$—SO$_3$H

                                                     |

                                                    CF$_3$

Figure 2.13 Examples of poly-fluorosulfonic acid-based polymers of short (a) and long (b) chain character, respectively.

facilitate the proton transport. The backbone provides mechanical stability for the PEM, and, due to the hydrophilic character of the side-chains, hydrophilic channels are developed within a hydrophobic matrix. The proton transport takes place via these channels.

Water molecules are needed for proton transport and are supplied by either water or humidified gases. As the PEMFC operates in the temperature range of 60–95° C, humidity can easily be controlled for high performance of the cell. At higher temperatures, the PEM exhibits high ion conductivity even at lower degrees of humidity.

The potential moves the protons from one water molecule to another within the aqueous channels. The thickness of the PEM affects the ion conductivity and therefore a thin PEM is preferred for increased cell efficiency. If the PEM is too thin, however, a decrease in cell performance can be observed due to gas and/or electrons moving in the opposite direction, i.e. from the positive to the negative electrode, a phenomenon commonly referred to as *crossover*.

Not only the protons, but also water moves in the hydrophilic channels, mainly due to electro-osmotic drag within the electric field between the negative and the positive electrodes. The water concentration along the PEM can vary and may result in a drying out of the negative electrode and flooding of the positive electrode. Back-diffusion of produced water at the positive electrode may occur, resulting in an improved water concentration gradient.

2.6.2   PEMFC usage

In order to fully utilise the benefits of the PEMFC technology, it is crucial to understand the factors limiting the performance. As for all electrochemical cells, degradation occurs due to usage, misuse, storage, or time. Therefore, some of the degradation mechanisms that affect the polarisation profile and limit cell performance will be discussed. Many of the degradation mechanisms, both physical and chemical, arise from the PEM, but also from the gases and the bipolar plates.

The PEM can be degraded by factors such as temperature, pressure, and water content. Small holes, *pinholes*, may be found in the PEM allowing gases to pass from one electrode to the other without taking part in the electrochemical reactions and, thus, resulting in capacity loss and, at worst, an internal short circuit. The pinholes may arise in the manufacturing process, by local dry-out of the PEM, or by freezing of the water inside the PEM. Dry-out of the membrane may also result in shrinkage of the membrane that consequently affects the mechanical stability both of the PEM and the electrodes. The temperature may increase locally in the PEM due to uneven reaction sites across the active area, creating *hot-spots*, which in the long run could create pinholes since the reaction rates are faster here generating even more heat to the hot-spots.

The chemical degradation of the PEMFC is mainly due to the acidic environment and side reactions. Radicals, such as hydroxyl or peroxyl (•OH and •OOH), may be formed at the positive electrode and react with the polymer to form HF, which in turn additionally degrades the polymer. Hydrogen peroxide may be produced at available catalytic sites at the positive electrode. Instead of forming water, the protons react with oxygen according to:

$$2H^+ + O_2 + 2e^- \rightarrow 2H_2O_2$$

If the bipolar plates are of metallic types, e.g. stainless steel, the radicals may also cause corrosion of the plates and dissolved metal ions could inhibit the proton conductivity in the PEM and thereby decrease overall cell performance.

Performance degradation may also originate at the electrodes. The main mechanisms are related to dissolution of the catalyst and clogging of the catalyst particles, resulting in, for example, structural changes and changes in the surface of the catalyst particles, and thereby less active catalytic sites. Moreover, the carbon support may corrode in the strong acidic environment.

As the main fuel for PEMFC is hydrogen, requiring either specific infrastructure or hydrogen generation onboard the vehicle. Fossil fuels like natural gas, gasoline, diesel, and other high-energy fuels may also be used in a fuel cell, but only at very high temperatures, e.g. in a solid oxide fuel cell (SOFC) (Table 2.2). For these fuels to be used in PEMFC, they must first be reformed to hydrogen:

$$C_xH_y + H_2O \rightarrow xCO + yH_2$$

The large amount of CO produced will, however, poison the Pt-catalysts, and the CO content is therefore preferably reduced according to the gas-shift reaction:

$$CO + H_2O \rightarrow CO_2 + H_2$$

This reaction also yields additional hydrogen. The power density will, however, decrease as the hydrogen gas contains CO_2 due to mass-transport limitations at the active catalytic sites. Even if the gas-shift reaction is exothermal, both reactions consume energy, decreasing the overall fuel cell efficiency. Impurities will also affect the fuel cell performance negatively, but can be eliminated partly or fully by different fuel cleaning procedures.

PEMFC and other fuel cell concepts can be used in vehicles: as primary propulsion power sources in all-electric vehicles or for auxiliaries, e.g. for air conditioning. As a primary power source, the PEMFC, not being suitable for fast and transient power responses, is often combined with a battery or an EDLC in order to achieve sufficient performance. Air must be supplied and hydrogen has to be stored or generated by a fuel reformer, and this clearly affects the overall power and energy density of the complete installation. Moreover, water and heat management must also be incorporated to secure optimal performance.

II

Li-ion battery technology – materials and cell design

3   Lithium battery materials

Among the different battery technologies possible for electric vehicles, the lithium-ion (Li-ion) concept stands out as the most attractive due to its combination of high cell voltage, high-energy density, and rate capability. The Li-ion battery is one of the members of the large family of lithium batteries, where the choice of different material combinations provides cells with different properties, each designed with a specific application in mind. Material requirements, possibilities and constraints, and how the materials effect energy and power characteristics will be the focus of the coming sections.

The rechargeable lithium battery families are categorised by the type of active material used as the *negative electrode*: metallic lithium or insertion[1]/conversion. If metallic lithium (often a thin metallic foil) is used, the result is either a *Li metal*, or a *Li (metal) polymer battery*, whereas insertion or conversion materials result in *Li-ion batteries*. Depending on the *electrolyte* used, either *Li-ion* or *Li-ion polymer* cells are achieved. The different families are summarised in Figure 3.1.

Apart from selecting the active material for the negative electrode, most often graphite, and the electrolyte, the selection of active material for the *positive electrode*

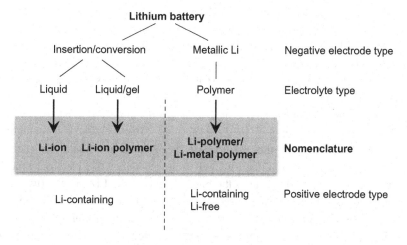

Figure 3.1 Different families of rechargeable lithium batteries.

---

[1] *Intercalation* is an often used term instead of insertion.

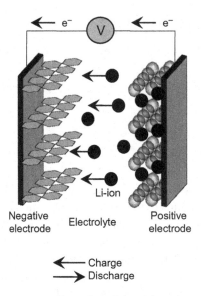

Figure 3.2 Illustration of the Li-ion concept.

will further define the performance characteristics of the cell, and some combinations are more preferable than others for electric vehicles. The active material for the positive electrode is of an insertion or conversion type. Often abbreviations like NCA and LFP cells are used, referring to Li-ion cells based on a negative electrode of graphite and a positive electrode of NCA or LFP (Sections 3.2.1.1 and 3.2.3).

The Li-ion battery is the most commonly used lithium battery in consumer electronics and electric vehicles, and the concept is illustrated in Figure 3.2. Although other concepts will be briefly explored, the Li-ion concept will be the model for further discussions of cell performance related to active and non-active materials.

During charge, the lithium ions are extracted from the positive electrode and transported via the electrolyte to the negative electrode where they are inserted. At the same time, the electrons pass via the external circuit. The reverse mechanisms take place in the discharge process.

The electrolyte comprises one or several *Li-salts* dissolved in one or several *solvents*. The general requirements are to allow easy salt solvation in the solvent and to facilitate ion transport between the electrodes. The electrolyte should be electrically and chemically stable within the full cell voltage range: the electrochemical stability window (Section 1.5.6). Due to the high voltage of the Li-ion cells (usually >2.5 V), non-aqueous solvents must be used; water decomposes above 1.23 V. Furthermore, in order to be of interest for vehicle applications, the electrolyte should have a low melting point and a high boiling point (i.e. a wide operational temperature range), and be non-toxic in the case of abuse situations. The types of electrolytes used in lithium batteries are *liquids*, *gels*, or *polymers*.

Many of the performance characteristics of a Li-ion cell depend on the nature of the active electrode materials, of both negative and positive electrodes, and their ability to reversibly insert and extract lithium. An attractive electrode material should exhibit high capacity and suitable potential vs. Li/Li^+, high for the positive electrode and low for the negative in order to achieve cells of high voltage, and to enable acceptable ionic and electronic conductivity. Moreover, the electrode materials should be compatible with the electrolyte and exhibit high Coulombic efficiency. Preferably they should also be non-toxic, and stable within a wide temperature range as well as in air, in order to simplify the production processes.

A vast number of materials may serve as electrode materials. Hence the chemistry in Figure 3.1 is not fixed, unlike the majority of non-Li battery technologies. Depending on the active materials selected, the cell can be produced in a charged or a discharged state, depending on where the cyclable lithium ions are located. The active material of the positive electrode in Li-ion cells must contain lithium in order to have access to cyclable lithium, and, thus, the cells are manufactured in a discharged state. On the other hand, if the negative electrode is metallic lithium, i.e. a Li-metal battery, a Li-free positive electrode is possible and this results in a charged cell as manufactured.

The electrodes of Li-ion cells are all *composite electrodes* (except for the special case of metallic Li) (Section 4.1). In addition to the active electrode material, often micro- and nano-sized particles, the composite electrode consists of the following inactive materials: a *binder* to hold the electrode particles together, and an *electron-conducting agent* (i.e. carbon black) to increase the electronic conductivity, all pasted on a *current collector*. These inactive materials are not directly involved in the electrochemical redox reactions, but they are nevertheless essential for overall electrode functionality. The current collector can be seen as the physical barrier of the cell, with the purpose to provide mechanical support and to simplify the electron transfer to the external circuit.

During cycling, when the active electrode materials incorporate lithium ions via insertion or conversion reactions, they also undergo volume changes. These volume changes should not affect the mechanical structure of either the active material particles or the composite electrode in order to maintain electrical contact within the electrode. Therefore, the composite electrode is normally provided with a porous structure in the electrode formulation and manufacturing. The thicknesses of the electrodes are design parameters optimising the cells for high-energy density or high-power density. A high-power-optimised cell (e.g. for HEV usage) will not necessarily contain the same active electrode materials as a cell designed for high energy (e.g. for an EV). Moreover, the surface area of the electrodes can be made very large to enable high current rates, especially for use in high-power applications.

A wide range of materials fulfil the above requirements. Depending on the Li battery family of interest, some material combinations are more attractive than others. Material selection is also influenced by the potential and the capacity of the materials. In Figure 3.3, many of the active electrode materials discussed are shown as potential versus theoretical capacity. For practical usage, material combinations providing a cell voltage of > 2.5 V should be chosen in order to obtain cells of reasonable energy density.

The performance of a cell does not only depend on the electrochemical properties, but also on how the different parts in the electrochemical cell interact, and how the electrodes and the cells are manufactured. Gaining control of cell performance through

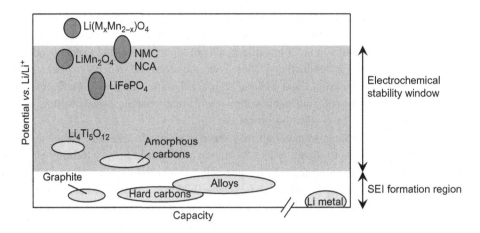

Figure 3.3 Potential and capacity of active electrode materials.

tailored material properties requires a fundamental understanding of the possibilities and limitations of all parts of the cell and the reactions and processes taking place at all levels. Moreover, many of the safety concerns arise from the material properties and have to be taken into account.

Material requirements, possibilities, and constraints will be described in the following sections. General as well as specific material properties will be discussed regarding both active materials and inactive materials, as separators and current collectors. As for the usability of electric vehicles, the energy and power content are important properties originating at the material level. Since these two properties are contradictory when it comes to cell design, they will be discussed both from a materials and cell perspective.

3.1   Negative electrode materials

There are two main types of active materials for the negative electrode: metallic lithium and insertion or conversion materials. Examples of insertion and conversion materials are graphite, alloys, and oxides. In Figure 3.4, some attractive materials are shown. The most commonly used material in Li-ion cells is graphite, mainly due to its low potential, only *ca.* 0.1 V higher than metallic lithium, and its ability to reversibly insert and extract a large amount of lithium ions.

In the following sections, selected active materials for the negative electrode will be discussed in terms of possibilities and limitations:

- metallic lithium
- carbon-based insertion materials: mainly graphite
- alloys: Si, SnSb
- oxides: conversion (MO_x) and insertion materials ($Li_4Ti_5O_{12}$)

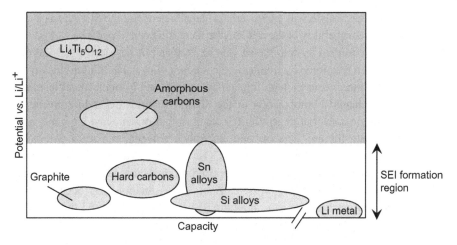

Figure 3.4 The potential and the capacity of some electrode materials related to the electrochemical stability window.

Before going into the details of the different materials, a fundamental prerequisite for many of them will be described: the stability towards the non-aqueous electrolyte. As can be seen in Figure 3.4, some electrode materials fall outside the electrochemical stability window, i.e. their potentials are above the LUMO level of the electrolyte, as described in Section 1.5.6. In order to function as electrode material, a protective layer to stabilise the electrode must be formed at the interface between the electrode surface and the electrolyte. The formation and properties of this *solid electrolyte interphase* (SEI) will be discussed next, before the specific characteristics of the various active negative electrode materials.

3.1.1   The solid electrolyte interphase

The formation of a *solid electrolyte interphase* (SEI) on the negative electrode surface in non-aqueous electrolytes is crucial to ensure performance and enable optimal usage of the Li-ion cell. The SEI layer is both a passivation and a protective layer. It is mainly formed from the decomposition products of the electrolyte, and it should cover the entire electrode surface in order to prevent any sustained electrolyte decomposition, which otherwise is favoured by thermodynamics. All active electrode materials operating outside the electrochemical stability window are in need of an SEI layer. The development of electrolytes that readily allow the formation of an SEI on the graphite surface was in fact the breakthrough enabling the commercialisation of Li-ion cells at the beginning of the 1990s.

Once formed, the SEI layer stays on the surface and acts like another phase or component within the cell, and is therefore recalled as an interphase (and not an interface). The main function of the SEI layer is to separate the negative electrode from the electrolyte and inhibit the electron transfer from the electrode surface to the solvent molecules of the electrolyte. The ion conductivity between the active electrode material and the electrolyte will, however, also be affected, and the SEI acts as a barrier limiting the power capability of the cell and the increasing cell impedance. A loss in cell capacity will be observed as both the solvent and the salt of the electrolyte take part in the SEI formation reactions, e.g. reducing the amount of cyclable lithium. Ideally, the SEI layer thus is thin, possesses high ionic conductivity, and is an electrical insulator.

The chemical composition of the SEI layer varies depending on the electrode and electrolyte used, but generally it is composed of organic and inorganic materials, often containing lithium. The organic electrolyte solvents can react with the salt and be decomposed into, for example, Li_2CO_3, CO, CO_2, ethylene, Li-oxalates, and Li-alkyl carbonates. As many of the decomposed products are gaseous, the pressure inside the cell will rise and in the worst case an abuse situation may occur. There are several proposed mechanisms as to how the SEI layer is formed, and the mechanisms behind the actual formation can be found elsewhere.[2]

The SEI layer may differ depending on the active material used, which can be illustrated by comparing the layer formed on a metallic lithium electrode with the one formed on a graphite composite electrode. Because of the similar potentials of fully lithiated graphite and metallic lithium (a difference of about 0.1 V), it would be possible to suppose that the two SEI layers are comparable. On the metallic lithium though, the SEI layer is formed as soon as the electrode comes in contact with the electrolyte. As for the graphite electrode, on the other hand, the SEI formation process takes place only under polarisation conditions because the intrinsic potential of the electrode is higher than the reduction potential of most salts and solvents, and therefore the SEI formation will proceed stepwise. A fundamental difference between a metallic lithium electrode and a graphite electrode influencing the SEI layer is the composite character of the latter where voids in the composite can accommodate both lithium ions and solvent molecules and therefore the SEI formation can take place not only at the electrode surface but also inside the composite electrode.

The thickness of the SEI varies depending on the electrodes and the electrolytes used, and is somewhere between a few atomic layers up to tenths of nanometres depending on, for example, the age and the charge state of the cell. Furthermore, the morphology of the layer is complex and changes with time, temperature, and electrolyte composition. The SEI can also be of a multi-layer character where lithium-rich components are found closer to the electrode surface, and components of a more organic character closer to the bulk of the electrolyte.

From an electric vehicle perspective though, it is essential to understand the nature of the SEI layer and how it affects cell performance over the entire life of the cell. The SEI layer influences the electrochemical performance of the electrode surface, such as thermal and chemical stability. Cycling, especially at elevated temperatures, can strongly affect the reactivity at the interface, and the SEI layer may grow or dissolve. Factors affecting SEI growth are, for example, the type of salt and solvent used, the nature of the electrode material, the electrode surface morphology, temperature,

[2] See e.g. P.B. Balbuena and Y. Wang (editors), *Lithium-Ion Batteries: Electrolyte Interphase*, World Scientific Publishing Company, 2004.

and impurities. The growth of the SEI layer can be examined using EIS measurements (Section 1.7.3), and the semi-circles in a Nyquist plot can be assigned to different SEI-related properties, where two semi-circles may arise: one at medium frequency corresponding to ionic migration processes in the SEI and one at a low frequency referring to lithium charge-transfer processes.

The mechanical strength of the SEI layer must also be adequate, in order to withstand expansions and contractions of the electrode during the charging and discharging processes, and additionally be well adhered to the electrode surface, and have low solubility in the electrolyte. Moreover, cracks can be formed in the SEI layer as a result of the volume changes occurring during cycling due to the insertion and extraction of lithium. Damaged SEI layers, regardless of cause, have to be re-formed by the same reactions that occurred in the initial SEI formation processes, and thus more cyclable lithium will be consumed, decreasing cell capacity accordingly.

During storage, not only temperature but also the SOC of the cell will affect the SEI layer, depending on, for example, storage time, temperature, and chemistry of the electrodes, as well as the composition of the electrolyte. For example, lithium ions may diffuse from the graphite structure through the non-uniform SEI layer and react with the electrolyte to form another SEI on top of the former. During storage, equilibrium would be reached between the SEI dissolution and Li diffusion rate. The net effect is loss of cyclable lithium and thereby an irreversible loss in cell capacity. The capacity loss is also due to the reactivity of the carbon surface, which is partially unprotected because of SEI dissolution. As a consequence, the parking conditions (e.g. time and temperature) of an electric vehicle may in the long run have a significant effect on the all-electric driving range and/or the life of the battery, depending on type of electric vehicle (e.g. EV or HEV, see Section 5.1.1).

Depending on the negative electrode used, the SEI formation can be optimised. As indicated in Figure 3.4, amorphous carbons are found within the electrochemical stability window and will not form an SEI layer. They can be used as a coating to optimise the surface properties of the electrode and thereby improve performance and reduce irreversible capacity loss. Artificial SEI layers can also be applied to further improve the performance of the negative electrode. The SEI layer may also be tailored by adding chemical compounds to the electrolyte: *additives*. The nature of these compounds and their role in the electrolyte, as well as in the cell, are described in Section 3.3.1.3.

3.1.2   Metallic lithium

The seemingly simplest way to introduce lithium into a battery is to use a metallic lithium negative electrode. Lithium is the lightest of all metals (0.54 kg/dm^3) and has a low electrochemical potential (-3.04 V vs. SHE). Like other alkali metals, lithium is very reactive to contact with air and moisture, and must be handled with care. The manufacturing processes of the cells should preferably be performed in moisture free environments. Despite this feature, the interest in lithium metal-based cells is still considerable due to the high theoretical capacity (about 3800 mAh/g). The melting point of pure metallic lithium is high, 180.5 °C, also enabling Li-battery concepts operating at elevated temperatures.

As a consequence of the low electrode potential of lithium, most solvents are unstable in the presence of metallic lithium, as indicated in Figure 3.4. The stability of metallic lithium in non-aqueous electrolytes is therefore due to the formation of a protective SEI layer on the electrode surface.

As assembled, a cell of metallic lithium has an excess of lithium in the negative electrode, and can be combined with either a lithium-containing or a lithium-free positive electrode, and is thus assembled in either a discharged or a charged state. This opens up possibilities for the usage of other types of positive electrode materials than for a Li-ion cell as Li-free materials, e.g. V_3O_8, V_6O_{13}, or MnO_2 can be used.

In the case of a lithium metal polymer battery, a thin layer of a polymer electrolyte is deposited on the positive composite electrode surface and assembled with the Li-metal negative electrode. The polymer electrolyte is stable against the negative electrode (i.e. a larger electrochemical stability window towards a lower potential) and no SEI layer will be formed on the electrode surface. Due to the compact cell construction, the resulting cell is of high-energy density, having low resistance and impedance. These cells operate over a wide temperature range, preferably between 60 °C and 90 °C. Depending on vehicle application, this temperature range can be an advantage since it may simplify the thermal management both of the battery and the complete vehicle; various components in the electric vehicle may share the same thermal management circuit to simplify the overall system complexity. The high operational temperature, however, requires the battery to be heated before any usage, which can be a drawback in some geographical regions, but by using smart vehicle integration and battery encapsulation the problem can be reduced.

The lithium metal electrode is of a blocking character (Figure 1.3) and the reactions take place at the electrode surface. During the charging process of the cell, lithium is deposited, or *plated*, back on to the metallic surface. Since the surface is not completely flat, the lithium metal deposition is not homogeneous, which results in spots of lower impedance or higher electric fields. This makes the surface rough and *dendrites* can be formed. During the subsequent discharge process, lithium is dissolved from the entire electrode surface, including the dendrites. During cycling, the surface area of the lithium metal increases significantly with a corresponding increase in the reactivity, affecting the stability of the cell. The dendrites may also be disconnected from the electrode and enter the electrode as highly reactive lithium clusters. The dendrite formation and the corresponding dissolution processes are schematically illustrated in Figure 3.5.

The dendrite formation is the primary failure mechanism of a Li-metal cell. The surface area of the lithium metal increases hugely with dendrite formation and rapid electrolyte decomposition may take place. The surface can, however, be stabilised in different ways: by using other types of electrolytes or salts, or by modifying the surface and applying different protective layers. The surface can also be stabilised by mechanical treatment in order to limit the nucleation sites for dendrite growth by making, for example, small holes in the surface.

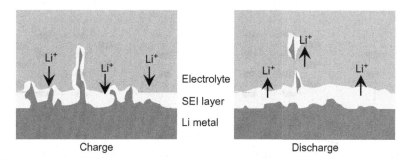

Figure 3.5 Schematic illustration of lithium dendrite formation and dissolution during charge and discharge.

3.1.3　Carbons

In the case of Li-ion cells, insertion materials made from carbon are the most common choice of active materials for the negative electrode. Soft carbons, hard carbons, and graphite are all suitable as active materials for the negative electrode of a Li-ion cell. Soft and hard carbons are amorphous, and graphite is crystalline, and the latter can be either synthetic or natural.

The reason for employing carbons rather than metallic lithium is mainly due to proven stability and production despite a lower capacity. The structure of the carbon material has a high influence on electrochemical performance, primarily both the capacity and rate capability. The macroscopic shape can be, for example, flakes, spheres, fibres, whiskers – all influenced by the production process and the source of raw material. For example, the orientation of the graphite particles affects the reversible capacity, where less oriented particles have a lower reversible capacity as a result of slower lithium insertion kinetics. The different types of carbon-based materials will be described in the following section, but since graphite is the material mainly used in the negative electrode of Li-ion cells, graphite will be the focus.

Graphite is a crystalline structure of hexagonal packed carbons in graphene layers held together by weak van der Waals forces. These layers are stacked in an ABA configuration, as illustrated in Figure 3.6. In the interstitial regions between the graphene layers, lithium can be accommodated to a maximum of one lithium atom per every six carbon atoms to form LiC_6, corresponding to a theoretical capacity of 372 mAh/g.

Lithium insertion into graphite is associated with a structural change. The ABA stacking sequence is transformed into an AAA order (Figure 3.6) and the lithium ions are inserted at distinct stages associated with two-phase voltage plateaus in the corresponding discharge profile. Each insertion stage also corresponds to a specific amount of lithium in the structure, where x in LiC_x varies in distinct steps: 6, 12, 18, . . ., as schematically indicated in Figure 3.7.

There are several models describing how the lithium insertion process takes place, and a classically used model is illustrated in Figure 3.8. The numbering of the stages corresponds to the number of graphene layers between the two lithium layers; i.e. the higher the stage number, the lower the amount of lithium in the structure.

(a)   (b)

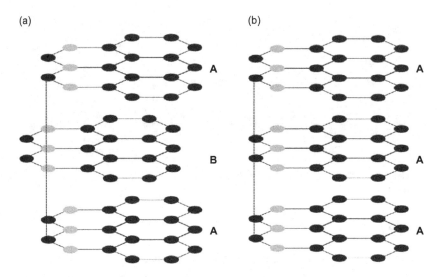

Figure 3.6 The graphite structure with (a) ABA and (b) AAA stacking.

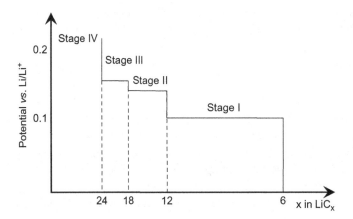

Figure 3.7 Schematic discharge profile of a graphite electrode indicating the different LiC_x stages.

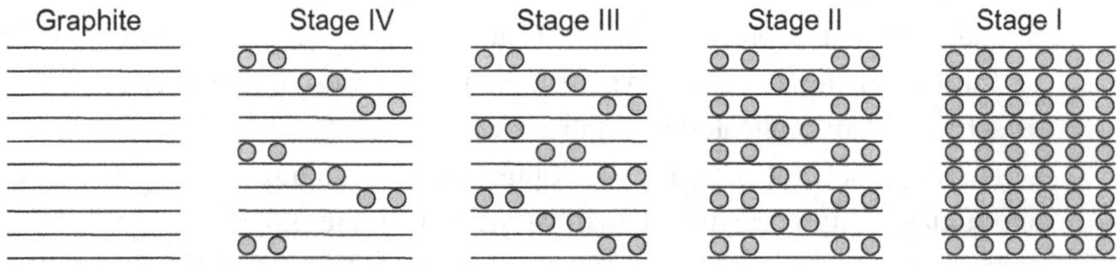

Graphite	Stage IV	Stage III	Stage II	Stage I

Figure 3.8 Model of lithium insertion into the graphite structure.

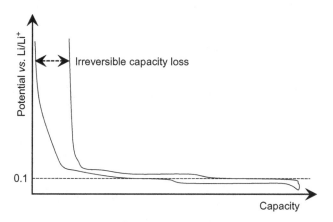

Figure 3.9 First charge-discharge of a half cell of graphite *versus* metallic lithium.

During the first insertion reaction of lithium into the graphite structure, i.e. the first charging process of the cell, some of the lithium ions react with the electrolyte to form the protective SEI layer at the electrode surface. The formation of the SEI layer causes an irreversible loss of cyclable lithium and hence a loss in capacity. Measurements using a graphite half cell versus a metallic lithium electrode can be used to visualise and quantify the irreversible capacity loss (Figure 3.9). As the charging and discharging processes continue, the subsequent discharge profiles follow the appearance of the second cycle.

The redox process of the graphite-based electrodes is influenced to a great extent by particle size, including specific surfaces, surface chemistry, morphology, crystallinity, and orientation of the crystallites. The stacking disorder of the graphene layers is significant for the insertion performance. The layers can still be parallel, but shifted or rotated, or they can be non-parallel: *turbostratic disordered graphite*. Completely random ordering is found in, for example, *carbon black*, a material commonly used for improving the electronic conductivity of the composite electrodes (Section 4.1). Lithium ion insertion is drastically reduced or inhibited between turbostratically disordered graphene layers. Therefore, the capacity depends on the degree of turbostraticallity, according to Capacity = $372(1-T)$, where T is the fraction of misaligned layers, and the fraction can be reduced by, for example, heat treatment.

As for cells using metallic lithium, the graphite negative electrode is made larger in capacity than the positive electrode in order to compensate for the capacity loss, but also in order to ensure that lithium is inserted into the graphite structure instead of being plated on top of the negative electrode. This latter phenomenon, *lithium plating*, is a side reaction within a Li-ion cell resulting in loss of cyclable lithium and thereby a loss in capacity. The underlying mechanism is due to the very small voltage difference between lithiated graphite and metallic lithium; about 0.1 V. On top of the plated lithium, a new SEI layer will be formed, according to the same mechanisms as for SEI formation on metallic lithium. Lithium plating in itself is reversible, as the plated lithium can be oxidised again. This, however, causes an overpotential during the discharge process,

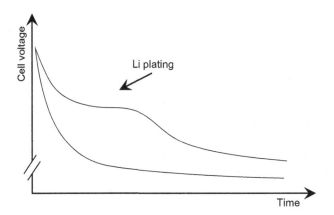

Figure 3.10 Discharge profile with and without lithium plating.

which can be observed in the discharge profile. An example is given in Figure 3.10. The plated lithium can, however, remain at the electrode surface in the case of unfavourable operating conditions, and will in such cases increase cell resistance.

There are several factors causing lithium to be plated on the graphite surface, e.g. the capacity ratio between the negative and the positive electrode (i.e. the N/P ratio, Section 4.3.1), the operating temperature, and the C-rate. These factors affect the kinetics of the negative electrode and the diffusion rate of the lithium ions so that lithium is plated on the surface rather than being inserted into the graphite structure. Moreover, the surface morphology, including the SEI surface morphology, may result in inhomogeneous current distribution along the surface, which has a considerable effect on the risk of Li-plating and the formation of dendrites, which, in the worst case, may cause an internal short circuit.

The Li-plating can be homogeneous or heterogeneous depending on the operating conditions and cell construction. In the homogeneous case, lithium is plated over a large area of the negative electrode and is mainly a result of cell charging at low temperatures. Heterogeneous Li-plating, on the other hand, occurs at non-uniform sites of the negative electrode where the current distribution is higher and is most often found at the edges and around impurity particles incorporated in the composite electrode. This type of Li-plating is typically related to the cell design and manufacturing processes, and defects originated therein. The heterogeneous Li-plating can be worsened by the conditions causing the homogeneous type. Figure 3.11 illustrates how Li-plating can be caused by an impurity particle at the electrode surface. It is therefore of key importance to avoid or limit impurities in the manufacturing process and sustain sound quality control of the entire production chain, from raw materials to final cell assembly. From a vehicle perspective, the control strategies have to include charge restrictions at low temperatures, often already at +10 °C, and restrict the charge rates accordingly.

The electrolyte also plays a crucial role for different failure modes of graphite electrodes. In the electrolyte, the lithium ions are surrounded by solvent molecules:

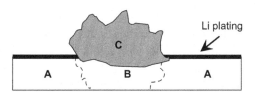

Figure 3.11 Example of Li-plating around an impurity particle at the negative electrode surface. A: Li excess regions, B: Li depleted region, C: impurity/high resistance region.

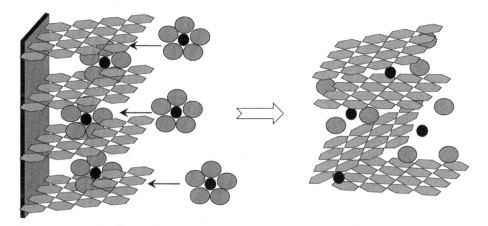

Figure 3.12 Schematic illustration of solvent co-insertion into the graphite structure.

the *solvation shell* $Li(solv)_x^+$. As the solvated lithium ions pass through the SEI layer, they lose the solvation shell which is inserted into the graphite structure as ions, as schematically illustrated in Figure 3.12. If the SEI layer is ineffective, *co-insertion* of solvent molecules can take place, i.e. the solvation shell will be intact during the insertion reaction. As the reaction proceeds, the solvent molecules will be reduced and gas may be formed (mainly CO and CO_2).

The gases formed may cause *exfoliation* of the graphene layers, damaging the structure and ruling out any further use of the graphite structure for lithium ion insertion/extraction. This destructive phenomenon is dependent on the solvent used in the electrolyte, and is mainly observed in electrolytes based on propylene carbonate. The graphite can be modified by thermal and mechanical treatment in order to minimise solvent co-insertion, and through these treatments the particles obtain a smooth surface with a reduced area. The smooth surface will reduce the exfoliation tendency of the graphene layers, and the reduced area results in a lower irreversible capacity loss. The mechanical treatments may also introduce structural defects supporting the formation of an effective SEI layer. Furthermore, the graphite electrode surface can be modified chemically by introducing functional groups, e.g. amines and carboxylic groups, reacting with the electrolyte components in order to enhance the stability of the SEI layer.

As indicated in the beginning of this section, amorphous carbons can also be used in Li-ion cells. Soft carbons are mechanically soft and can be produced from a variety of carbon sources. They can be transformed into graphite upon heating (about 2000–3000 °C).

Hard carbons, on the other hand, cannot be transformed into graphite at any temperature, and are strongly disordered materials. Hard carbons exhibit high capacity since the random alignment of the graphene layers provides significant porosity capable of accommodating lithium. As the lithium ions are inserted between the graphene layers, hard carbons can incorporate lithium fast. Due to the random orientation of the graphite particles, lithium ions can be trapped in the amorphous carbons, resulting in an irreversible capacity loss.

3.1.4   Alloys

At room temperature, lithium can electrochemically alloy with a number of metals, e.g. Al, Si, Sb, Mg, Zn, by simple and reversible redox reactions:

$$x\text{Li}^+ + xe^- + \text{M} \leftrightarrow \text{Li}_x\text{M}$$

These binary lithium-rich Li_xM materials are appealing due to their very high theoretical capacity. The alloying potential (vs. Li/Li^+) is higher than the insertion voltage of graphite (Figure 3.4), but still lower than the electrochemical stability window of the electrolyte, requiring the formation of an SEI layer.

Silicon is one of the most attractive metals due to its high theoretical capacity (4212 mAh/g fully lithiated, $\text{Li}_{22}\text{Si}_5$) and its abundance in the Earth's crust. Lithium will alloy with silicon at about 0.1–0.6 V vs. Li/Li^+ according to two-phase behaviour, i.e. a flat discharge profile is observed. The first cycle is associated with high irreversible capacity loss due to the transformation from a crystalline Si phase to an amorphous phase, and the formation of a protective SEI layer. Therefore, maximum capacity corresponding to the composition of $\text{Li}_{15}\text{Si}_4$ (theoretically *ca.* 3600 mAh/g) can be reached only if extremely low C-rates are used, about C/100 or lower – rates insufficient for electric vehicle applications (often C/2–10C is required, Section 5.1), and, consequently, practical capacity will be much lower. In addition, the irreversible loss will continue during subsequent cycles and at a faster rate, resulting in even lower capacity.

As an active material for negative electrodes in Li-ion cells, the Si-based electrode is made as a composite, in the same manner as the carbon-based electrodes. When Si forms alloys with lithium, a significant volume change will occur; an increase of about 300% from the pure Si phase. This volume expansion and contraction upon cycling will induce high mechanical stress in the material, which may lead to degradation of the material and severe cracking, resulting in high and rapid capacity losses. Moreover, the volume changes can result in loss of electrical contacts within the composite electrode, as well as with the current collector. The loss of contact between particles results in lithium being trapped in the active Si particles, and consequently there is a loss of cyclable lithium. Moreover, the SEI layer will also be affected by the large volume changes and may crack during the charging and discharging processes. The large volume changes during alloying with lithium are also seen for other metals.

There are ways to minimise and circumvent the issues related to volume changes; for example, by minimising the active particle size and using nano-sized Si, and carefully constructing the composite electrodes, including selection of a binder material, or utilising thin films. Using amorphous phases of the alloying metals can also be an alternative. These electrodes will, however, be of lower capacity due to the limited available sites for lithium compared to the corresponding crystalline phase, even if the cycling performance can be just as good or even better.

Another strategy is to create a micro-structured composite comprising an active Li-alloy phase uniformly embedded in an inert host matrix, such as Si/C or Si/TiN. This will enhance the cycle life since the host matrix functions as a buffer to accommodate the volume changes. The matrix will also increase the electrical contacts between the particles and minimise the mechanical stress of the electrode.

The amount of matrix material used will of course lead to a trade-off between volume changes and the loss of specific capacity, as the addition of an inactive material to the electrode will increase the weight. The matrix can, however, also be of active character, inserting lithium and thus increasing the capacity, as long as the volume changes are accommodated. For example, graphitised carbon can be used as buffer for volume expansions, and at the same time insert lithium ions alongside the Li-Si alloying reaction. Combining the cycling stability of graphite with the high capacity of Si will equip the negative electrode with new properties while the main capacity will still be due to the Li-Si alloy. Other examples of composite electrodes are based on amorphous carbon, intermetallic materials, and metal oxides.

Furthermore, alloys in the form of intermetallic materials can be used as the active electrode material, e.g. SnSb, Cu_6Sn_5, and Cu_2Sb. The redox reaction route and the reaction with lithium can be either of insertion, formation, or conversion types (Section 1.5.5.1). The insertion reactions are associated with small effects on the crystal structure and small volume changes. These materials exhibit good reversibility, but a limited amount of inserted lithium, i.e. low capacity.

The formation and conversion reactions, on the other hand, involve full reduction of a transition metal to its metallic state, thus enabling materials of high capacity. These reactions are associated with structural rearrangements and large volume changes between the delithiated and the lithiated phases and thereby low cycling stability, even if high capacity can be achieved. The metals of the intermetallic materials can be either active or inactive *vis-à-vis* the lithium reaction, according to the following examples:

$$SnSb + Li \leftrightarrow Li_{22}Sn_5 + Li_3Sb \text{ (both metals are active)}$$

$$Cu_6Sn_5 + Li \leftrightarrow Li_{22}Sn_5 + 6Cu \text{ (only one metal is active)}$$

Some of the intermetallic materials can react further with lithium in one or two steps. For example, the reaction of SnSb above is a two-step reaction according to:

$$\text{Step 1: } SnSb + 3Li^+ + 3e^- \leftrightarrow Li_3Sb + Sn \text{ (extrusion reaction)}$$

$$\text{Step 2: } 5Sn + 22Li^+ + 22e^- \leftrightarrow Li_{22}Sn_5 \text{ (formation reaction)}$$

One drawback of conversion reaction electrodes is the large voltage hysteresis (Section 1.6.1.1), limiting the energy efficiency of the cell. The hysteresis is related to the high polarisations needed to break the intermetallic bonds.

3.1.5   Oxides

As will be described in more detail later, in the context of positive electrode materials, oxides can readily incorporate lithium and they also enable different kinds of pathways for fast lithium diffusion. Oxides are thus one of the main groups of materials attractive for the positive electrode, but can also serve as the active material of the negative electrode. There are mainly two types of oxides used for the negative electrodes: conversion and insertion oxides.

The *conversion oxides* are often used as a complement to the alloys/intermetallic materials described above, and are used to buffer the volume changes during lithiation/delithiation processes. Examples of such oxides are SnO, SnO_2, SiO_2, PbO, ZnO, Fe_3O_4, Co_3O_4, and Al_2O_3 and they react with lithium in a two-step process according to:

$$\text{Step 1: } MO_x + 2xLi^+ + 2xe^- \rightarrow M + xLi_2O$$

$$\text{Step 2: } M + yLi^+ + ye^- \leftrightarrow Li_yM$$

The Li_2O formed acts as a buffer for the volume changes associated with the reversible reaction in Step 2. The reaction in Step 1 is, however, often associated with large irreversible capacity loss, but, depending on the M used, it can be reversible. As two examples, Co-based oxides are reversible and Sn-based oxides irreversible.

Just as in the case of lithium alloying materials, nano-sized materials will provide a strain accommodation for the lithiation/delithiation processes and will improve the electrochemical performance due to the larger electrode/electrolyte interface improving the kinetics and enabling short electron and ion conduction paths. The increased electrode/electrolyte interface can, however, result in unwanted degradation of the electrolyte, due to the catalytic nature of the nano-sized particles.

3.1.5.1 $Li_4Ti_5O_{12}$

The oxides of *insertion type* are capable of inserting lithium reversibly and usually have high operational potential compared to metallic lithium or graphite, and are therefore often attractive as active materials of the positive electrode. There are, however, a limited number of alternatives suitable for the negative electrode as well: the titanium oxide, $Li_4Ti_5O_{12}$ (LTO), is the premium example, and it will be used to illustrate this type of active material for the negative electrode.

LTO crystallises in the cubic spinel structure (space group Fd3m) and is iso-structural to $LiMn_2O_4$ (an active material for positive electrodes, Section 3.2.2). The spinel structure enables three-dimensional lithium ion diffusion pathways and the volume of LTO is stable, almost unaffected, upon cycling. Indeed, during the lithiation process the structure of LTO is reversibly transformed to a rock salt structure, caused by the lithium ordering within the host structure, but it results in a negligible volume change (about 0.2%): LTO is therefore often named a zero-strain material and has a long cycle life with stable mechanical integration of the composite electrode.

The LTO material has a potential of approximately 1.55 V vs. Li/Li$^+$ and reacts with lithium according to:

$$Li(Li_{1/3}Ti_{5/3})O_4 + Li^+ + e^- \leftrightarrow Li_2(Li_{1/3}Ti_{5/3})O_4$$

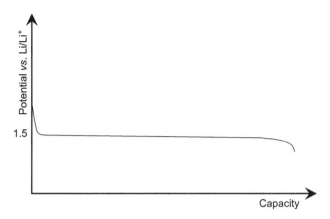

Figure 3.13 Schematic discharge profile for the LTO material.

The electrochemical reaction is of two-phase character forming solid solutions of the two phases, resulting in a flat discharge profile, almost ideal as schematically shown in Figure 3.13. The theoretical capacity is 175 mAh/g, about half of the capacity of graphite. The usable capacity of the LTO material is still lower, limited to cycling of one out of two lithium ions in the structure since Ti cannot be further oxidised.

The higher potential of 1.55 V vs. Li/Li$^+$ results in lower cell voltage than for a corresponding cell with a graphite electrode, and the energy density will consequently be lower as well and more cells will be required in the battery compared to using cells of higher voltage. The large advantage of the higher potential of LTO is that the electrode always stays within the electrochemical stability window of the electrolyte (Figure 3.4), and thus no SEI layer of decomposed electrode materials is formed on the electrode surface.

Moreover, the absence of an SEI layer when using LTO as the negative electrode material opens up the possibility to use other types of electrolytes (which cannot be used in combination with graphite) with, for example, improved low-temperature perform-ance, which may result in cells even more attractive for electric vehicles.

Furthermore, due to the high operational potential of the negative electrode, the risk of lithium plating is eliminated, which is a safety advantage. Another advantage of the LTO material is the possibility to use aluminium as the current collector, which will result in lighter cells and further improvements in energy density.

The LTO material as made, with micro-sized particles, has limited electrical conduct-ivity and low rate capabilities. This is due to limited lithium ions kinetics within the structure, and therefore surface or size modifications are employed, e.g. carbon coatings or the use of nano-sized particles. This way electrical conductivity is enhanced, and in combination with the fast 3D diffusion pathway network for lithium within the crystal structure the rate capability becomes excellent. In combination with a positive electrode of LiMn$_2$O$_4$, LTO is therefore a highly attractive alternative for power-optimised cells to be used in, for example, HEVs.

The thermal stability of LTO, up to *ca.* 200° C, is a safety advantage as is the lack of a sensitive SEI layer. There are some drawbacks of this material though; gases such as H_2, CO_2, CO, CH_4, may evolve at the charged state of the cell, i.e. in the presence of Li_2 ($Li_{1/3}Ti_{5/3}$)O_4, and they become distinct at elevated temperatures (above *ca.* 45 °C) mainly due to the catalytic activity of the surface decomposing the electrolyte. Optimising the electrode/electrolyte interface and improving electrolyte composition may limit gas evolution, whereas lithium impurities, such as Li_2CO_3 and LiOH, in combination with an insulation layer on the electrode surface, can accelerate gas evolution.

3.2   Positive electrode materials

The active materials of the positive electrode for Li-ion batteries operate by reversibly incorporating lithium in the structure, preferably without any significant structural changes in the host material. This secures long life and high efficiency. There is no ideal material that meets all the requirements for all types of applications, which is why a large variety of materials have been developed in order to tailor cell performance. In Li-ion cells, the negative electrode and the electrolyte are most often the same in all types of cells, i.e. graphite and the 'standard electrolyte' (1M $LiPF_6$ in EC/DMC). Therefore, the positive electrode is the main selection to be made and crucial for overall cell performance and the characteristics of the electric vehicle. The selection influences the all-electric range, acceleration performance, temperature dependency, and durability.

Active electrode materials for the positive electrode can be either lithium containing or lithium-free. The former is most commonly used, and the latter requires a negative electrode including cyclable lithium, e.g. metallic lithium. The active material must be able to reversibly insert/extract a large amount of lithium in order to achieve a cell of high capacity. To achieve high-power rates, the insertion/extraction processes must be fast, with high kinetics and high ion conductivity within the material, as well as at the electrode/electrolyte interface.

The positive electrode is normally made starting from an insertion material having at least one reducible/oxidisable species; materials containing one or several transition metals in order to handle the inserted lithium ions by a change in the oxidation states of the transition metals. In Figure 3.14, the potential versus the capacity of some relevant materials are shown. In many cases, as in the case of electric vehicles, high electrode potential is desirable, since this results in a cell of high voltage and large energy density.

As can be seen from Figure 3.14, positive electrode materials fall into two main categories: oxides and polyanionics of transition metals depending on the chemical and electrochemical potentials (equations (1.5)–(1.8), Section 1.4.1) of the materials. The chemical potential of the material should match the HOMO level of the electrolyte, i.e. the potential should be within the electrochemical stability window. The electrochemical potential is related to the oxidation state of the transition metal(s), and will change as a function of the amount of lithium ions inserted into the active material. Charge compensation during lithium insertion/extraction is enabled by redox reactions, which will affect the electronic state of the material.

The requirement for high specific capacity generally restricts the choice to materials containing first-row transition metals, e.g. manganese, cobalt, iron, and nickel, due to

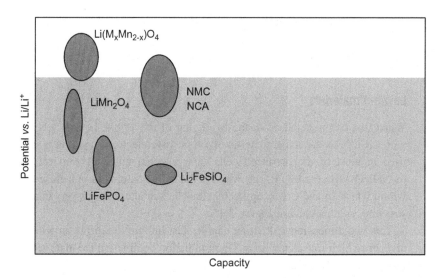

Figure 3.14 The potential as a function of the capacity of commonly used active positive electrode materials.

the weight of the elements needing to be minimised. Often the active materials contain a mixture of transition metals to improve, for example, electrochemical performance or capacity, or to increase cell voltage. Materials able to reversibly undergo more than one oxidation state change per transition metal could substantially increase capacity. In this respect, vanadium and chromium offer wide and attractive ranges of changes in the oxidation states, but they have both been precluded in the discussion due to environmental and toxicity issues.

The sites available for lithium atoms in the material structure (e.g. octahedral or tetrahedral sites), and how they are linked, will affect the lithium diffusion pathways within the solid host structure. The sites can be linked in one, two, or three dimensions, consequently creating differences in potentials, rate capabilities, and stabilities.

At potentials above *ca.* 4.2 V vs. Li/Li$^+$, a *surface protective layer* (SPL) of decomposed electrolyte products is formed on the positive electrode, much like the SEI layer on the negative electrodes. This layer will increase the polarisation resistance of the cell, and increased impedance will be observed. To overcome the formation of these degradation products, a surface coating of an inorganic material, e.g. AlF_3 and Al_2O_3, can be applied to the electrode particle surface. This will improve the rechargeability of the cell.

In the next sub-sections, specific materials suitable as positive electrode materials will be presented. Only materials applicable to the Li-ion and Li-ion polymer concepts suitable for electric vehicles will be included, but the general terms can be used also for a proper understanding of lithium-free active positive electrode materials for the lithium-polymer concept. The following materials will be discussed in terms of crystal structures and type of polyanion used:

- layered: $LiCoO_2$, $LiNiO_2$, and derivatives thereof
- cubic spinel: $LiMn_2O_4$

- orthorhombic: $LiFePO_4$
- polyanionic: e.g. $Li_2Fe_2SiO_4$
- mixed materials

3.2.1   Layered materials

When first commercialised in the beginning of the 1990s, Li-ion cells were equipped with $LiCoO_2$ as the active material of the positive electrode and this is still the material used in most of the produced cells. The material has a layered rock-salt structure (α-NaFeO$_2$, space group R-3m) wherein the oxygen atoms form a closed-packed lattice within which Li and Co occupy octahedral sites in alternating layers in a sandwich-like structure, as illustrated in Figure 3.15.

This two-dimensional structure enables fast lithium ion diffusion within the material and hence high-rate capabilities. Upon lithiation/delithiation the material is a one-phase insertion type material:

$$xLi^+ + xe^- + Li_xCoO_2 \leftrightarrow LiCoO_2$$

During charge and discharge, the oxidation state of Co varies between III and IV. The charging is, however, limited to approximately 4.2 V vs. Li/Li$^+$, corresponding to about 0.5–0.6 Li cycles due to phase transitions with strong mechanical strain at higher potentials, where the structure would collapse and hinder further lithium insertion/ extraction. The theoretical capacity is 274 mAh/g (i.e. extraction of all lithium from the structure), but the practical capacity is hence limited to about 150 mAh/g.

To overcome the structural instability and increase the amount of extractable lithium (i.e. increase the capacity) several alternative layered materials based on the $LiCoO_2$ structure have been developed. These layered structures are often referred to as the 'LiMO$_2$' structure, where M is a single transition metal or a mixture of several transition metals.

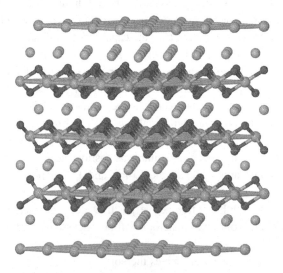

Figure 3.15 The layered $LiCoO_2$ structure.

One such alternative material is LiNiO$_2$, possessing a usable capacity of about 170–200 mAh/g related to the redox couple Ni^{III}/Ni^{IV}. The synthesis routes are, however, complicated and a phase-pure material is not always achieved. As for LiCoO$_2$, the Li and the Ni occupy interstitial sites in the layers between the oxygen lattices. The material has, however, a strong tendency towards cation mixing both in the Li layer and the Ni layer, where Ni atoms may block the Li diffusion and thereby reduce the usable capacity; Ni^{II} is oxidised to Ni^{III} in the Li layers and at the same time the lattice contracts and consequently the distance between the oxygen sandwich layers decreases. Not only is the lithium ion diffusion limited, the space where lithium ions enter also decreases resulting in increased insertion polarisation and irreversible loss of capacity.

LiNiO$_2$ is, however, both thermally and structurally unstable. The material will decompose and oxygen will be released reacting with the solvents of the electrolyte through exothermal reaction, and the corresponding reduction of the transition metals is associated with a structural change, from 2D to 3D, at the surface of the active electrode particles. The structure can, however, be stabilised by Co substitution on the Ni sites, even if thermal instability remains. A common degree of substitution is 80% Ni and 20% Co: LiNi$_{0.8}$Co$_{0.2}$O$_2$. This material is, however, sensitive to both air and water, and a surface layer of Li$_2$CO$_3$ is formed when in contact with air and the material delithiates chemically in contact with water, even more pronounced than without the Co substitution. This makes the material difficult to handle during the manufacturing process – the active electrode material as such – and in the manufacturing of the composite electrodes and the cell assembly process.

The layered material can be even further improved by introducing other substitutions in the structure: aluminium for the NCA material and manganese for the NMC material, both described below.

Despite all efforts to stabilise the structure, thermal stability is still an issue for layered 'LiMO$_2$' materials. Figure 3.16 shows the results of differential scanning calorimetry experiments performed to follow the heat evaluation during the

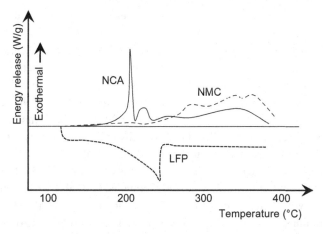

Figure 3.16 The thermal reactions of the decomposition of NCA, NMC, and LFP obtained by differential scanning calorimetry experiments.

decomposition reactions occurring at elevated temperatures. For comparison, the LFP material (Section 3.2.3) is shown in Figure 3.16. As can be seen, the exothermal reactions of the NCA material are quite drastic, while the LFP material undergoes endothermal reactions, which also destroy the material, but do not cause any abuse situations.

3.2.1.1 $LiNi_{0.8}Co_{0.15}Al_{0.05}O_2$

To further improve the layered structure, a common strategy is to substitute both by Al and Co at the Ni sites. The most common material is $LiNi_{0.8}Co_{0.15}Al_{0.05}O_2$, also known as NCA. Aluminium improves the thermal stability and the electrochemical perform-ance, and Co limits the cation mixing in the Ni layer, resulting in a cell of improved cycling performance. The enhanced 2D structure enables the material to cycle a large number of lithium atoms without collapsing over a large number of cycles, resulting in long-lasting cells.

This stabilisation of the structure is due to the modifications of the neighbourhood in the crystal structure around the transition metals. The Ni cations are responsible for the redox reactions by the redox couple Ni^{III}/Ni^{IV}. The Co^{III} imposes the 2D character of the structure, whereas the Al^{III} delays the 2D to 3D transition at elevated temperatures. The exothermal reactions cannot, however, be fully eliminated, and the capacity of the cell highly influences the reaction pathway. Despite the modest thermal stability, the NCA material is still very attractive for electric vehicle applications due to the power performance and high capacity (about 180–200 mAh/g).

3.2.1.2 $LiNi_{1/3}Mn_{1/3}Co_{1/3}O_2$

Manganese can also be used to substitute Ni in the $LiNiO_2$ structure to further stabilise the structure during charging/discharging. $LiNi_{1/2}Mn_{1/2}O_2$ is a stable material and Mn is attractive compared to, for example, Al, because of its ability to change the oxidation state and thereby improve capacity and enable higher potential. At discharged state, Ni^{III} and Mn^{III} coexist, and at charged state Mn is oxidised to Mn^{IV} while Ni is reduced to Ni^{II}. The presence of Mn^{III} in the structure will, however, make the structure unstable. This phenomenon will be further described in connection to $LiMn_2O_4$ in Section 3.2.2. Due to the oxidation state of the Mn, this material is difficult to synthesise, and therefore Co is also used as a substitution element to even further stabilise the structure, resulting in the material $LiNi_{1/3}Mn_{1/3}Co_{1/3}O_2$, known as the NMC material, or the '1/3, 1/3, 1/3' material. In the discharged state, the oxidation states of the metals are Ni^{II}, Co^{III}, and Mn^{IV}, and the stability is thus improved by the lack of Mn^{III}. The composition of an NMC material used in a marketed cell is not necessarily exactly the same as the original NMC material; the composition may vary from '1/3, 1/3, 1/3'.

The NMC material can be charged to higher potentials than $LiCoO_2$ and $LiNiO_2$, and the material is a further step towards improved thermal stability due to the stabilisation effects of Mn^{IV}. Even if the theoretical capacity is about 280 mAh/g, the practical capacity is still in the same range as that of $LiCoO_2$ and $LiNiO_2$ (170 mAh/g).

The voltage profile of the NMC material is of a sloping character and the changes in the oxidation states of the transition metals used are related to specific parts of the

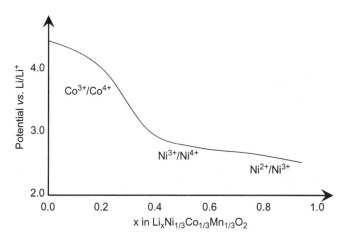

Figure 3.17 The discharge profile of the NMC material and the corresponding changes in oxidation states for the transition metals.

discharge profile, as schematically shown in Figure 3.17. During discharge, starting from the high-potential region (> 4 V vs. Li/Li$^+$), the potential decrease is associated with the change in the oxidation state of Co from Co^{IV} to Co^{III}, and corresponds to an insertion of 1/3 of the lithium ions. Further insertion of lithium relates to changes in the oxidation state of Ni from Ni^{IV} to Ni^{III}, in the region of 1/3 to 2/3 of inserted lithium, and for the insertion of the remaining 1/3 of the lithium ions, Ni is further reduced from Ni^{III} to Ni^{II}.

It is possible to further increase the capacity of NMC by structurally integrating Li-rich NMC with layered Li$_2$MnO$_3$. The resulting material is both Li and Mn rich, and can be referred to as zLi$_2$MnO$_3$ – (1-z)LiNMC, where the overall Li to transition metal ratio is > 1. The Li$_2$MnO$_3$ material is by itself, from a structural point of view, electrochemically inactive, but can be activated at potentials above 4.5 V vs. Li/Li$^+$. In total, high-energy density is achieved, either as composites containing the two materials or as a solid solution between the two, and the synthesis procedure is critical for the final material. The material is stable and enables fast lithium ion diffusion. These layered high-voltage materials are attractive from an electric vehicle perspective due to the capacity (>230 mAh/g), but the materials do suffer from voltage hysteresis proportional to the Li$_2$MnO$_3$ content, which causes a decline in cell voltage upon cycling, and consequently a decline in the energy density of the cell.

3.2.2 The cubic spinel LiMn$_2$O$_4$

The cubic spinel LiMn$_2$O$_4$ material (space group Fd3m) has the general structure of the AB$_2$O$_4$ type; a cubic close-packed oxygen lattice wherein A and B occupy tetrahedral and octahedral sites, which is illustrated in Figure 3.18. The cubic spinel structure is very stable and frequently found in nature as the gem MgAl$_2$O$_4$.

As a lithium insertion material, LiMn$_2$O$_4$ is the basic configuration where a mix of Mn^{III} and Mn^{IV} occupies the octahedral sites and Li occupies the tetrahedral sites,

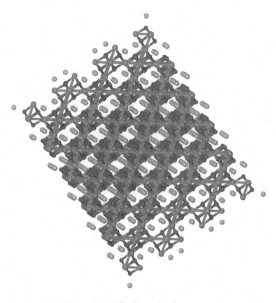

Figure 3.18 The cubic AB_2O_4 spinel type structure.

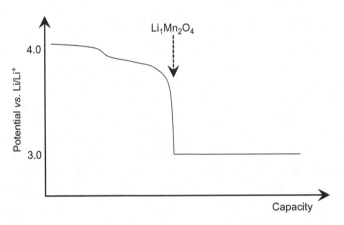

Figure 3.19 Voltage profile of the $LiMn_2O_4$ material.

resulting in a three-dimensional lattice of lithium diffusion pathways within the structure, enabling high rate capabilities. The theoretical capacity is 148 mAh/g, and the practical capacity is about 100–120 mAh/g. The material is still of interest for electric vehicles, despite the comparatively low capacity, mainly due to cycling stability and rate capability.

A schematic discharge profile for the $LiMn_2O_4$ material is shown in Figure 3.19. The profile has two main plateaus, at *ca.* 4.0 V and at 3.0 V, corresponding to Mn^{IV} and Mn^{III}, respectively. Theoretically, the cell can be either charged or discharged and cycled over a wider range $Li_xMn_2O_4$ $0 \leq x \leq 2$ to give a theoretical capacity of 285 mAh/g.

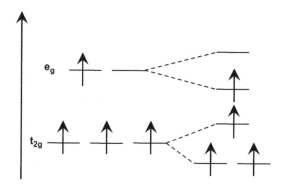

Figure 3.20 The Jahn–Teller distortion effect of the Mn^{III} orbitals and symmetry.

The insertion of lithium to form $Li_2Mn_2O_4$, however, creates a distortion of the crystal structure, accompanied by a structural change from cubic spinel to tetragonal, with a corresponding volume expansion (16%). This is caused by a Jahn–Teller distortion due to the surplus of Mn^{III}. The Jahn–Teller distortion is associated with specific electron configurations of some metal elements, such as Mn, resulting in changes to the geometrical configuration around the atoms. In crystalline materials, the distortion is often observed in octahedral complexes and is enhanced if an odd number of electrons occupy e_g orbitals since a distortion will lower the energy as the octahedral complex becomes tetragonal. The Mn^{III}, in its high-spin state, has three electrons in t_{2g} orbitals, and one in e_g, and thereby undergoes a Jahn–Teller distortion, as illustrated in Figure 3.20. The distortion will take place only as the Mn^{III}/Mn^{IV} ratio is > 1, i.e. Mn^{III} is dominant in the structure.

These structural changes make it practically impossible to insert further lithium into the structure beyond $Li_1Mn_2O_4$ since the mechanical stress in the active material particles and the composite electrode become too large. The consequences are that the electrical contacts particle-to-particle or with the current collectors may be lost. Therefore, only the upper plateau in the discharge profile of Figure 3.19 is used, where the cubic structure expands/contracts symmetrically. At very fast discharge rates, however, a minor amount of $Li_2Mn_2O_4$ can be formed at the surface of the active material particles.

There are ways to decrease the Mn^{III}/Mn^{IV} ratio by substitution with other metals, thereby limiting the Jahn–Teller distortion. Even if the capacity of the material is derived from the Mn^{III}/Mn^{IV} ratio, substituted materials can exhibit high capacity and enhance cycling performance characteristics. One example is Li-substituted materials, $Li_{1+x}Mn_{2-x}O_4$, which give lower capacity at the 4 V plateau, but the structure is more stable at even higher potentials. At $x=1/3$, the maximum degree of substitution is achieved; all Mn are now in the Mn^{IV} state and no further extraction of lithium can take place. Another method to increase the performance of the material is, for example, to introduce cation vacancies in the structural lattice to decrease the Mn^{III}/Mn^{IV} ratio, resulting in new potential pathways for lithium ion diffusion. The degree of vacancies is, however, difficult to control in the production process.

At elevated temperatures or at low SOC levels, capacity fading is observed for $LiMn_2O_4$. Manganese can be dissolved from the structure affecting the structural stability at high and low potentials. The mechanism proceeds according to:

$$2Mn^{III} \rightarrow Mn^{IV} + Mn^{2+}$$

and is often referred to as the *Hunter's disproportional reaction*. The Mn^{2+} ions are dissolved in the electrolyte, and this loss of Mn will result in a defect spinel structure with limited capacity due to an increased amount of Mn^{IV} in the structure, limiting the lithium extraction. The disproportional reaction is emphasised if traces of H_2O or H^+ are present in the electrolyte, and the following total reaction will take place:

$$2LiMn^{III}Mn^{IV}O_4 + 4H^+ \rightarrow 3 \; \lambda\text{-}Mn^{IV}O_2 + Mn^{2+} + 2Li^+ + 2H_2O$$

The dissolved Mn^{2+} may migrate towards the negative electrolyte to block the lithium diffusion pathways and thereby even further reduce the amount of cyclable lithium. Moreover, at the negative electrode, the Mn^{2+} ions can be reduced and precipitate as metallic Mn, which will further increase the charge-transfer resistance at the interface between the negative electrode and the electrolyte. The Mn^{2+} may also form impurities in the SEI layer of graphite negative electrodes, which could cause side reactions. Hard carbon-based electrodes, on the other hand, are not that sensitive to Mn^{2+}. Adding a surface layer, e.g. a thin film of Al_2O_3, on the $LiMn_2O_4$ composite electrode can minimise the Mn^{2+} dissolution.

3.2.2.1   High-voltage spinel alternatives

The efforts to minimise the Mn^{III}/Mn^{IV} ratio can result in materials of higher potentials. This will enable the use of negative electrode materials of higher potentials, and still result in a reasonable cell voltage. This will furthermore also reduce the issues associated with the formation of the SEI layer when carbon-based negative electrodes are employed.

Spinel structured materials with high potentials, close to 5 V vs. Li/Li^+, can be achieved by substitution of some of the Mn with other transition metals, usually Ni, Cu, Fe, Cr, or Co. The resulting materials all exhibit two voltage plateaus: at about 4 V corresponding to the Mn^{III}/Mn^{IV} redox couple, and at about 5 V corresponding to the change in the oxidation state for the dopant metal: Ni^{II}/Ni^{IV}, Cr^{III}/Cr^{VI}, etc. The capacity of the 4 V plateau depends on the Mn^{III}/Mn^{IV} ratio. Table 3.1 summarises some of the key parameters for some examples of these types of high-potential materials.

Table 3.1 Key parameters for high-potential spinel structured materials.

Material	Charge distribution	Redox couple	Potential vs. Li/Li^+
$LiNi_{0.5}Mn_{1.5}O_4$	$Li[Ni^{II}_{0.5}Mn^{IV}_{1.5}]O_4$	Ni^{II}/Ni^{IV}	4.7
$LiCo_{0.5}Mn_{1.5}O_4$	$Li[Co^{III}_{0.5}Mn^{III}_{0.5}Mn^{IV}]O_4$	Co^{III}/Co^{IV}, Mn^{III}/Mn^{IV}	5.0
$LiCoMnO_4$	$Li[Co^{III}Mn^{IV}]O_4$	Co^{III}/Co^{IV}	5.0
$LiFe_{0.5}Mn_{1.5}O_4$	$Li[Fe^{III}_{0.5}Mn^{III}_{0.5}Mn^{IV}]O_4$	Fe^{III}/Fe^{IV}, Mn^{III}/Mn^{IV}	4.9
$LiCr_{0.5}Mn_{1.5}O_4$	$Li[Cr^{III}_{0.5}Mn^{III}_{0.5}Mn^{IV}]O_4$	Cr^{III}/Cr^{IV}, Mn^{III}/Mn^{IV}	4.8

The stability of the electrolyte at high voltages is the key issue for the usage of these kinds of positive electrode materials. Therefore, new electrolyte formulations are often needed and furthermore also other parts of the composite electrodes may be subjected to degradation processes at high voltages. For example, the aluminium current collector, the binder material, or the electric-conducting particles may all oxidise, resulting in side reactions.

3.2.3   Olivine LiFePO$_4$

So far, materials with 2D and 3D character and lithium diffusion pathways have been described, and they have all been oxides. Now it is time for a material of polyanionic character with 1D diffusion pathways: LiFePO$_4$ (in short LFP). LFP crystallises in the orthorhombic olivine structure (space group Pnma) (Figure 3.21) where Li and Fe are located in octahedral sites and P in tetrahedral sites in a distorted hexagonal closed-packed oxygen array. Lithium diffusion takes place via 1D channels formed in the structure. Therefore, phase-pure materials are preferred since defects like impurities may totally block the diffusion pathways.

The theoretical capacity of LFP is intermediate (170 mAh/g) and the rate capability is poor, even at low C-rates, due to the limited electronic conductivity of the lithiated and delithiated phases. Therefore, the active material LFP particles must be surface-coated by, for example, amorphous carbon in order to achieve composite electrodes of acceptable rate capabilities. Moreover, utilising nano-sized particles will increase the electronic conductivity due to shorter diffusion distances. The electrochemical properties are also improved by the carbon coating, dependent on both the amount and the type of carbon. Moreover, a large surface area is required to improve the solid-state kinetics.

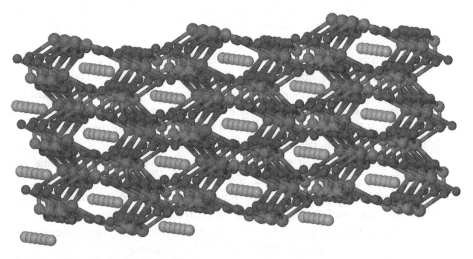

Figure 3.21 The orthorhombic LiFePO$_4$ structure.

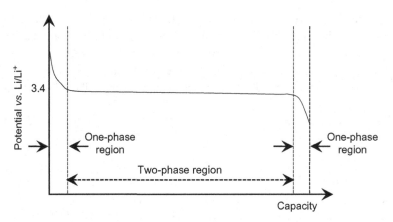

Figure 3.22 The voltage profile of LiFePO$_4$.

The LFP material has a potential of *ca.* 3.45 V vs. Li/Li$^+$ independent of the Li-content, i.e. a two-phase reaction having a flat discharge profile (Figure 3.22). It is based on the following reaction where the iron changes oxidation state from Fe^{II} to Fe^{III}:

$$LiFe^{II}PO_4 \leftrightarrow Fe^{III}PO_4 + Li^+ + e^-$$

At very low or very high lithium contents, i.e. almost fully charged or fully discharged, a single-phase character is observed as sloping parts in the voltage profile. At the fully charged state, the delithiated FePO$_4$ phase exists and as lithium starts to be inserted during the discharge process, a solid solution of Li$_\alpha$FePO$_4$ starts to form and the voltage drops accordingly. On the other hand, at the fully discharged state, only the fully lithiated LiFePO$_4$ phase is present. As the extraction of lithium takes place, at the beginning of the charge process, LiFePO$_4$ is present as the solid solution Li$_{1-\beta}$FePO$_4$ and the voltage increases accordingly. In the two-phase region in between, at the voltage plateau, the two solid solutions co-exist in different ratios, depending on the progress of the charge and discharge processes.

There are several theories, or models, developed to explain the two-phase mechanism occurring at equilibrium, i.e. at the voltage plateau, and two of them will be described here: the *core-shell model* and the *many-particle model*, both illustrated in Figure 3.23. In both methods, there is an equilibrium between the two phases: the lithium-rich phase Li$_{1-\beta}$FePO$_4$ and the lithium-poor Li$_\alpha$FePO$_4$. Moreover, both methods can equally be used to explain the different phenomena observed.

In the core-shell model, the surface of the particle becomes lithiated during the discharging process. As the discharging proceeds, the lithiated shell of the particle grows at the expense of the lithium-free core, and the interface between the two decreases until the fully lithiated LiFePO$_4$ phase has been reached for the entire particle. The opposite procedure takes place during the subsequent charging process; the surface of the particle becomes delithiated at the same time as lithium ions diffuse from the core to the surface.

(a)

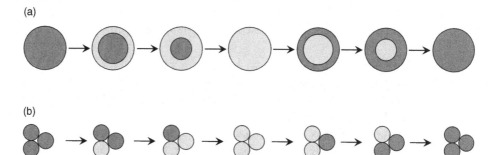

(b)

Figure 3.23 The reaction mechanisms in LiFePO₄ according to (a) the core-shell model and (b) the many-particle model.

In the many-particle model, on the other hand, the discharge process leads to lithiation particle by particle until all particles are fully discharged and during the subsequent charging process, the particles are delithiated one by one until the fully charged state has been reached.

The thermal stability of the LFP material both at charged and discharged state is high. If the delithiated FePO₄ is exposed to high temperatures (*ca.* 350 °C), an irreversible phase-transition takes place without any release of O_2 gas, in stark contrast to the case of the layered NMC and NCA materials. This lack of oxygen release is mainly due to the oxygen atoms being strongly bound in the phosphate anion unit of LFP. Disadvantages of the LiFePO₄ material are low energy density due to the low potential vs. Li/Li⁺, and low packaging efficiency of the composite electrode making the material difficult to use in energy-optimised cells.

There are other possible active electrode materials equivalent to LiFePO₄ based on PO₄ but on other transition metals; LiMPO₄ where M is, for example, Mn or Co, or mixed with Fe. The LiMnPO₄ material has a potential of about 0.5 V higher vs. Li/Li⁺ than LFP, which results in a specific energy density increase of about 15%. An even higher potential is achieved by LiCoPO₄: 4.8 V vs. Li/Li⁺, resulting in about 40% higher energy density. The rate capabilities of these two materials are poor, however, and LiMnPO₄ is unstable when charged (due to the Jahn–Teller effect of the Mn[III] dominance in the structure). Moreover, the ionic and electronic conductivities both of LiMnPO₄ and LiCoPO₄ are considerably lower than those of LFP, making the composite electrodes much less energy dense.

3.2.4   Other materials

There are other positive electrode materials that may be of interest in an electric vehicle, e.g. polyanionic materials of the form $Li_xM_2(XO_4)_3$. M refers to a transition metal and X is, for example, P or S, or a mixture thereof. In these materials, M is octahedrally coordinated and X tetrahedrally, in a lattice corner-sharing tetrahedral XO₄ and octahedral MO₆. Depending on the combination of M and X, materials can be tailored for specific electrochemical properties. Some combinations, however, result in non-rechargeable materials, i.e. only of interest for primary batteries.

In the following sections, two groups of materials that do have some promise with respect to EV usage will be described: Li_2MSiO_4 and $LiMS/PO_4F$.

3.2.4.1 Li_2MSiO_4

These insertion materials are based on silicate chemistry and are promising active electrode materials: Li_2MSiO_4, M=Fe, Mn, Co or a mixture thereof. Within the structure Li, M, and Si occupy tetrahedral sites within a distorted close-packed oxygen array. Networks of connected interstitial sites form tunnels for lithium diffusion within the structure. Generally, these materials exhibit high-temperature stability due to the strongly bound oxygen atoms in the SiO_4 polyanion, but the overall performance is very dependent on the production process.

The capacity of an active electrode material can be increased by: (i) increasing the average electrode potential, (ii) increasing the number of electrons involved in the redox reactions, and (iii) decreasing the molecular weight per mole electrons exchanged. The Li_2MSiO_4 materials are of interest due to their (in theory) ability to electrochemically extract two lithium ions, i.e. double the capacity – via (ii) above.

One example is Li_2FeSiO_4 with a theoretical capacity of about 330 mAh/g. Lithium is extracted during the first charging process at *ca.* 3.1 V vs. Li/Li$^+$, but the subsequent Li insertion and extraction reactions take place at 2.8 V, indicating a phase transition during the first charging process. The usable capacity is, however, about 130 mAh/g, much lower than the theoretical, but still comparable with, for example, the $LiMn_2O_4$ spinel. This strongly indicates that in practice less than one lithium ion can be reversibly inserted/extracted. According to the discharge profile of Li_2FeSiO_4, no further extraction of lithium can take place, i.e. Fe cannot be further oxidised. It is possible to extract more lithium from the structure and thereby increase the capacity by means of substitution with other transition metals. The most obvious choice is Mn by virtue of its Mn^{II} and Mn^{IV} oxidation states according to:

$$Li_2(Fe_{1-x}{}^{II}Mn_x{}^{II})SiO_4 \leftrightarrow Li_{1-x}(Fe_{1-x}{}^{III}Mn_x{}^{IV})SiO_4 + (1+x)Li^+ + (1+x)e^-$$

This material is attractive due to sustainability, and the abundance both of Fe and Mn. A substitution of about 20% results in theoretical capacity exceeding 200 mAh/g. A very schematic discharge profile of the $Li_2Fe_{1-x}Mn_xSiO_4$ material vs. Li/Li$^+$ is shown in Figure 3.24.

3.2.4.2   Metal fluorides

Most oxides and polyanionic materials rely on insertion reaction mechanisms. Metal fluoride materials, on the other hand, are based on using the conversion reaction mechanism. This is a group of positive electrode materials having higher energy densities than the corresponding oxides due to the high ionic character of the M-F bond. Mainly two types of metal fluorides will be discussed: *fluorophosphates* ($LiMPO_4F$) and *fluorosulphates* ($LiMSO_4F$), both often with potentials higher than 4.5 V vs. Li/Li$^+$. Transition metals of *3d* character are favoured: M=V, Co, Fe, Ti, Mn, etc. Even if the molecular weight of the framework units are higher, thus resulting in reduced energy density, the polyanionic materials exhibit very stable framework

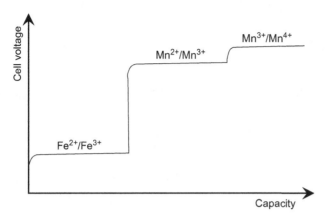

Figure 3.24 Schematic discharge profile for $Li_2Fe_{1-x}Mn_xSiO_4$.

structures and a wide range of substitution possibilities. The high electrode potential and also fast lithium diffusion, especially in the case of M=V are advantages, while one general shortcoming is low electric conductivities. The main drawback, however, is the stability of conventional electrolytes at these high potentials.

An example of an electrochemically active material that can be used either as a positive *or* negative electrode is $LiVPO_4F$. This material exhibits two plateaus: at 1.8 V and 4.2 V vs. Li/Li^+, resulting in a cell voltage of 2.4 V. At the plateau at 1.8 V, lithium is inserted into the $LiVPO_4F$ structure, associated with the V^{III}/V^{II} redox couple, and at 4.2 V lithium is extracted towards VPO_4F, involving the V^{III}/V^{IV} redox couple. The reversible capacities at both plateaus are similar and in the range of 150 mAh/g.

3.2.5   Mixed electrode concepts

Depending on specific performance or safety constraints, the positive electrode can be made by combining two or several active materials. This can be achieved either by mixing particles of different materials or by mixing different materials in one particle. The latter is often referred to as a 'core-shell' approach or compositionally graded materials. Mixed materials can be very attractive, from a pure performance perspective and furthermore also for the simplified SOC estimation, of significant importance for batteries of electric vehicles (Section 6.2.1).

The materials used in the combinations can have divergent electrochemical or safety properties; they might have different capacity or thermal stability, or operate at different voltage levels. For example, NMC or NCA can be used in the particle core to ensure high capacity, while a material exhibiting thermal stability can be used at the surface, creating a synergistic composite material. Likewise, mixing the $LiMn_2O_4$ spinel with NMC or NCA can limit the Mn dissolution from the spinel structure, by trapping H^+ or Mn^{2+} within the layered oxides, while the spinel material enhances thermal stability.

As mentioned above, besides improved performance, mixed materials can also facilitate an easier SOC estimation, a key factor in electric vehicle operations.

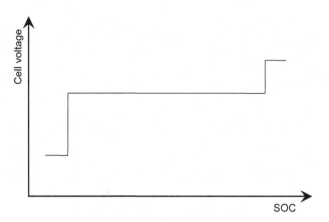

Figure 3.25 Schematic illustration of a mixed electrode for SOC control.

If the original potential profile is flat, as for $LiFePO_4$ or Li_2FeSiO_4, a small amount (< 10 wt%) of another electrochemically active material exhibiting two-phase behaviour with distinct potential steps can be used to more easily detect when a specific SOC level has been reached, as is schematically illustrated in Figure 3.25. This way, the SOC estimation can be made precise and the calibrations of several battery control algorithms can be adjusted during operation or on a regular basis, leading to much more flexible control, which is at the same time less complex.

3.3   Electrolytes and separators

While the combination of electrodes determines cell voltage, capacity, and energy of the cell, other important features are determined by the electrolyte, e.g. the rate at which the energy can be released (or accepted). The main purpose of the electrolyte is to facilitate lithium ion transport between the electrodes and to hold the electrodes apart to inhibit short circuits. The electrolyte should furthermore not take part in any redox reactions, and it could therefore be seen as inert, but is still a very dynamic component of the cell. For a successful cell to be accomplished, the electrolyte must be carefully selected to meet the specific demands of the electrode chemistry, as well as the operational requirements.

The electrolyte should be electrically and chemically stable and be of insulating character to avoid self-discharge and internal short circuits. The solvents and the salts of the electrolyte should ideally be non-toxic and environmentally friendly, and the electrolyte would preferably have a wide thermal stability region, a low melting point, and a high boiling point. In the ideal case, the electrolyte should exhibit high stability against the electrodes, which will be more and more difficult to maintain since positive electrode materials of high potentials are pursued, and improved electrolytes are therefore required. The electrolyte is sandwiched between the negative and the positive electrodes, and also partly integrated with them; the interfaces formed are essential for stable cell performance. Stability should be also maintained against other parts of the cell, such as separators, current collectors, and packaging material. In most situations, however, the stability is not that ideal. During charge and discharge, the electrodes may increase or decrease in volume due to the lithium content in the active materials. These volume changes are often taken care of by the design of the composite electrodes, but the electrolyte can also act as a buffer in order to keep the cell dimensions intact during cycling.

The vast majority of all electrolytes consist of Li-salts dissolved in one or several solvents. In lithium cells, non-aqueous solvents must be used due to the operating potentials; water decomposes to H_2 and O_2 above 1.23 V vs. Li/Li$^+$. Different types of electrolytes can be used depending on performance requirements and the operational parameters of the cell. In cells suitable for electric vehicle applications, liquid organic electrolytes are the most commonly used and will therefore be the focus of this chapter. In the following sections, the different parts of liquid electrolytes will be described, including solvents, salt, additives, and separators. The fundamentals of alternative electrolyte concepts – polymer electrolyte and ionic liquids – will also be discussed along with their advantages, limitations, and possible usage.

A basic electrolyte cannot be optimal for all applications as they can have very different cell requirements. It is, however, possible to optimise the performance of a basic electrolyte for specific applications and demands by including *additives* in the formulation. Only using some form of standard electrolyte is not recommended due to the many decomposition reactions possible. Additives can, for example, limit or contribute to the formation of the SEI layers, to reduce the electrolyte consumed and/ or to improve the ion conductivity across the interfaces between the electrodes and the electrolyte. The latter results in improved fast-charging and high-power capabilities. Moreover, many of the electrolyte solvents are flammable. This is especially problem-atic for the large-sized cells applicable for electric vehicles as the safety risk is often proportional to the size (in Ah) of the cell. By additives it is possible to create flame-retarded or even non-flammable electrolytes, which are thus of large interest to the vehicle industry.

3.3.1   Liquid electrolytes

The most commonly used electrolytes in Li-ion cells are liquid and this type of electrolyte and its constituents will therefore be described in more detail in the following sections.

3.3.1.1   Solvents

To meet the basic requirements of an electrolyte, the solvents should fulfil some basic characteristics. They should be able to dissolve salts at the desired concentrations, i.e. the solvents should have high dielectric constants and/or donor numbers, and in order to enable fast ion transport, the solvents should exhibit low viscosity to increase the bulk conductivity, according to equations (1.21)–(1.23), Section 1.5.4. To be safe and versatile they should possess high chemical and electrochemical stability against the other cell components. The solvents may also play a crucial role in the insertion/ extraction processes in Li-ion cells, e.g. the risk of detrimental co-insertion of the solvents into graphite-based negative electrodes.

Due to the strongly reducing and oxidising environments at the negative and the positive electrodes, the solvents must be free of active protons, as both oxidation and reduction of H^+ would take place within the operational voltage range of a lithium cell. To be able to dissolve lithium salt, the solvents need other polar groups, e.g. carbonyl (C=O), nitrile (C≡N), sulphonyl (S=O), or ether links (−O−). The majority of solvents are esters, ethers, or carbonates, e.g. ethylene carbonate (EC), propylene carbonate (PC), and dimethyl carbonate (DMC). Most electrolytes include mixtures of two or more solvents, while single-solvent electrolytes are uncommon in Li-ion cells. The reason for using several solvents is the diverse and often contradicting requirements depending on cell usage, which seldom can be met by electrolytes based on a single solvent. The lithium salts, on the other hand, are rarely mixed, due to the limited number of possible anions.

Ethylene carbonate, the most commonly used solvent, has a melting point of about 36 °C, and to extend the operational temperature range, EC is often mixed with other organic compounds, preferably of linear carbonate types. DMC is one of the most commonly used *co-solvents*, and the mixture can be stable up to potentials around 5 V vs. Li/Li$^+$. In the case of an EC/DMC solvent mixture, the EC content is usually lower than 50 vol%, preferably lower than 30 vol%, since higher concentrations cause precipitation at low temperatures. The properties of the EC/DMC mixture are dependent on the ratio and the temperature, as illustrated by the binary phase diagram in Figure 3.26.

The eutectic point in Figure 3.26 is typical of binary solvent mixtures, where both components mutually dissolve in the liquid state and do not dissolve in the solid state.

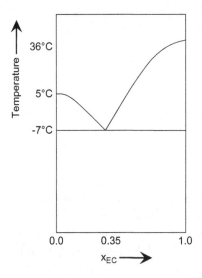

Figure 3.26 Binary phase diagram of mixtures of EC and DMC.

To ensure a wide operational temperature range, the composition near the eutectic point is crucial and is preferably expanded towards low temperatures. When adding a Li-salt, and several additives, to create a functional electrolyte, the thermal behaviour will, however, be different.

Below −20 °C, EC precipitates and drastically reduces the conductivity, and above +50 °C the solvent and the Li-salt will react vividly together and furthermore the Li-salt may act as a catalyst for further reactions. The low-temperature performance limitations are possible to overcome as soon as the cell has reached higher temperatures (usually above +10 °C), but the reactions taking place at the elevated temperatures result in irreversible performance losses. The reaction products are often gaseous, which might lead to abuse situations due to the pressure increase inside the cell. A broader operational temperature range is desired for electric vehicles to secure full performance and to limit abuse situations, e.g. vehicles parked in hot climates.

3.3.1.2 Salts

The salt must also fulfil some basic requirements in order to be of interest for Li-ion cells. It should be completely dissolvable and have a high transference number of lithium ions, t_{Li+}, as discussed in Section 1.5.4, in order to achieve high capacity and power capability. Moreover, the dissolved salt should not form crystalline solvated phases within the electrolyte, especially at low temperatures. The choice of anion affects to a considerable extent the conductivity and ion–ion interactions, such as ion pairing. Furthermore, the anion should withstand the highly oxidative and reductive environment and must not react with, for example, the solvent, the electrodes, or the separators at any operational temperatures. Salts of very high purity are preferred in order to minimise side reactions.

The number of potential salts suitable for Li-batteries is limited as many of the required properties are dependent on the overall electrolyte formulation, i.e. solvents, salt concentration, and additives. One of the main challenges is the fact that many Li-salts are not soluble in non-aqueous solvents. Lithium halides, for example, are impossible to use due to low solubility and/or instability of the anions at high potentials. The salts of interest are therefore based on more complex anions, and the most studied salts are $LiClO_4$, $LiBF_4$, $LiAsF_6$, $LiCF_3SO_3$ (Li-triflate), $LiN(SO_2CF_3)_2$ (LiTFSI), and $LiPF_6$, where the latter is the salt found in almost all commercial Li-ion cells.

The $LiPF_6$ salt has by far the best balance of the properties necessary for Li-ion cells, while each of the other salts listed above might be more suitable with respect to a few properties. In a mixed solvent electrolyte, e.g. based on EC/DMC, the $LiPF_6$, salt is stable at high potentials, about 5 V vs. Li/Li^+, which makes it very attractive.

There are, however, drawbacks of $LiPF_6$ not to be neglected, mainly concerning the performance and safety of a Li-ion cell. $LiPF_6$ is always in equilibrium with the gaseous PF_5 via the following reaction:

$$LiPF_6 \text{ (s)} \leftrightarrow LiF \text{ (s)} + PF_5 \text{ (g)}$$

This equilibrium is present at room temperature and even more favoured at elevated temperatures. PF_5 is a strong Lewis acid, and in non-aqueous solvents it tends to initiate side reactions. Moreover, both $LiPF_6$ and PF_5 are very sensitive to water, even traces of moisture, and react according to:

$$LiPF_6 + H_2O \leftrightarrow LiF + 2HF + POF_3$$

$$PF_5 + H_2O \leftrightarrow 2HF + POF_3$$

where HF can further react with the solvent, the SEI layer, and/or the electrodes. The LiF formed can be incorporated in the SEI layer, which will grow even further, causing increased cell impedance and a decreased amount of cyclable lithium. Despite these severe drawbacks, $LiPF_6$ is still used, but careful handling must be secured, especially at cell assembly, to limit the moisture traces in the cell.

Both Li-triflate and LiTFSI are promising candidates for Li-ion cells due to their high thermal stability, even if the ion conductivities of the corresponding electrolytes are lower than, for example, the $LiPF_6$ based, and even if they have mainly been used in combination with polymer electrolytes. The main drawback is the inability of these salts to passivate the Al current collector. This can, however, be eliminated using highly concentrated electrolytes (see below) or using ionic liquids as the solvent (Section 3.3.4).

The *salt concentration* in the electrolyte is a crucial parameter. As discussed in Section 1.5.4, the salt concentration affects the ion conductivity and is a trade-off mainly between the number of charge carriers and the ion mobility. At low salt concentrations, the number of charge carriers limits the conductivity and at high concentrations ion pairing is a strongly contributing process that might reduce conductivity significantly. In general, the maximum ion conductivity is therefore found in intermediate salt concentrations, balancing these two limitations. The ion conductivity is also dependent on the solvent used and its interaction with the salt. A commonly used salt concentration is 1 M of salt in an EC-based solvent, which is simply a trade off between the different parameters affecting the ion conductivity. The ion conductivity is, however, not the only factor influencing cell performance; the transference number is equally important (Section 1.5.4), and a high transference number will facilitate fast kinetics due to a reduction of the polarisation at the surface of the electrode. Depending on the combination of solvents and salts, high salt concentrations (i.e. >3 M) can be utilised in order to increase insertion kinetics mainly due to the high concentration of ions near the electrode surface, which is of capacitive character.

3.3.1.3   Additives

As for the negative and positive electrodes, it is almost impossible to find an optimal electrolyte based only on salt and solvent(s) serving all possible electrode combinations and applications. A commercial electrolyte for Li-ion cells therefore usually contains a multitude of additives, with the exact composition of the electrolyte differing among manufacturers; exact recipes are often secret. An *additive* is a chemical compound added in small amounts to the electrolyte in order to achieve optimal performance and to tailor the electrolyte for specific applications. The use of additives will, however, always create a more complex electrolyte.

The additives should be stable in the electrolyte and in the electrochemical environment at all operational temperatures, and they should not add to electrolyte viscosity, nor coordinate lithium ions, or take part in the main redox reactions of the cell, i.e. they should not limit the capacity and the power capability of the cell. Since the additives are present only in small amounts (about 5 wt%), they have only a minor/negligible impact on the overall weight and volume of the cell.

There are many types of additives stabilising and improving different performance characteristics. Below the main functionality of some additives applicable for performance and safety issues in electric vehicles will be described: *SEI forming, redox-shuttle, shut-down*, and *flame-retardant additives*.

SEI forming additives aim to control the thickness and stability of the SEI layer. During cycling, the SEI layer on the electrode surface may grow but the thickness of this layer can be controlled by deliberately causing additive decomposition to form a controlled and thinner SEI layer. There are a number of unique additives suppressing decomposition – this is achieved by causing these additives to undergo decomposition before the main electrolyte solvents decompose. Examples of SEI-forming additives are compounds containing unsaturated carbon–carbon bonds, such as vinylene carbonate (VC) and oxalates, and sulphur-containing compounds, such as cyclic sulphonates, sulphites, or sulphates. Moreover, VC can also suppress the co-insertion of solvents and prevent exfoliation of graphite. VC is reduced at the graphite negative electrode at potentials above 1 V vs. Li/Li^+.

The usage of specific overcharge/overdischarge limiting additives, *redox-shuttle additives*, can decrease the electrochemical sensitivity of Li-ion cells. Overcharge is a critical risk arising during operation. During overcharge, lithium is forced to leave the positive electrode and is not handled by insertion/conversion mechanisms at the negative electrode. If graphite is used as the negative electrode, this lithium may be plated on the graphite surface instead of being inserted into the graphite structure (Section 3.1.3), often resulting in irreversible degradation of the cell and corresponding loss of capacity.

As the amount of extractable lithium is depleting at the positive electrode, other chemical degradation processes are triggered, such as structural decomposition and gas release. Side reactions are used by additives to passively balance the cell by absorbing current without increasing the state of charge of the cell. These redox-shuttle additives are often used as a reversible overcharge/overdischarge protection device inside the cell. The chemicals are oxidised at the positive electrode and form positively charged radicals that diffuse via the electrolyte to the negative electrode where they are reduced. This mechanism effectively shuttles the charge away without causing any irreversible degradation reactions, e.g. Li-plating. The challenge, though, is finding additives with these reversible redox reaction properties.

Furthermore, as the redox shuttle reactions are voltage dependent, different additives must be used depending on the cell chemistry used. The redox potential of the additive should preferably be slightly higher than that of the positive electrode. A redox shuttle cannot prevent all abuse situations, though, since only a limited amount of charge can be shuttled. Iodine is one example of a redox-shuttle additive of Li-ion cells having a redox potential of 3.25 V vs. Li/Li$^+$ associated with the following reaction:

$$3I^- + 2e^- \leftrightarrow I_3^-$$

There are several types of *shut-down additives* all aiming to stop any further redox reactions, and often acting as irreversible overcharge protection. The general requirement of any shut-down additive is that the oxidation potential should be lower than the oxidation potential of the electrolyte, but higher than the voltage corresponding to a fully charged cell, and lower than the abuse overcharge voltage. Thus the shut-down additives are activated at specific voltages and therefore different additives are required depending on overall cell chemistry and operational voltage ranges. Moreover, the additives must have a high oxidation rate in order to act fast enough. Two main types will be presented here: *film-forming* and *gas formation additives*. The film-forming additives, mostly based on aromatic organic compounds, will form a thick insulating polymeric layer on the electrode surface to block any further lithium extraction or insertion. If a gas-forming additive is used, large amounts of gas will be generated in overcharge conditions thus increasing the internal cell pressure, which, in turn, might trigger a *current interruption device* (CID, Section 4.4.4) to deactivate the cell. The drawback of these additives is the total irreversibility of the reactions, making the cells useless.

Due to the high flammability of the organic electrolyte solvents, abuse situations can occur if the cell casing is damaged. *Flame-retardant additives* are chemicals suppressing continued combustion after the cell has been exposed to heat, sparks, or flames. Radical scavenging stops chain reactions in the gas phase during the combustion. By forming an insulating layer between the electrode and the gas phase, the additive will inhibit any further combustion reactions. Moreover, the flame-retardant additives can act to reduce the self-heating rate of the electrolyte. Examples of flame-retardant additives for Li-ion cells are organic compounds containing phosphorus and halogens, or ionic liquids.

Additives can be counterproductive and interfere with each other; they may enhance one property, but hamper another. Understanding how the different additives interact with each other is therefore essential. Moreover, additives always have side effects, which must be kept under control. Therefore, other additives are added to minimise those adverse reactions. In practice, several additives are often included in the electrolyte, working together in order to optimise the same target function. Having one additive addressing several functions would be beneficial or finding the synergetic effects of additives arising through combined use of specific single functional additives. The difficulty of optimising functional electrolytes for specific applications and usage conditions requires deep knowledge about additives, cell design, and how the usage conditions affect the cell. Furthermore, additives may lose their functionality and might degrade upon storage, especially at elevated temperatures.

3.3.2   Separators

The separator is a non-active component of the cell used together with liquid electrolytes. Its physical properties influence to a considerable extent overall cell performance, reliability, and safety. The main functions are to hold a large amount of electrolyte and to separate the electrodes. The separator is typically an electrically insulating porous membrane, with a thickness preferably uniform over time to ensure the same performance throughout the entire cell life. It should also be chemically and electrochemically stable towards the electrolyte and the electrodes. The separator affects cell impedance, operating temperature range, charge and discharge rates, and cycle-life. Moreover, the separator distributes the electrolyte evenly over the electrode surfaces as a result of its morphology and permeability.

Ideally, the separator should be as thin and light as possible to limit the non-active volume and mass of the cell. A thin separator with preserved mechanical strength is desired for fast ion transport and to achieve high-energy and power densities of the cell. Thinner separators are desirable in vehicle applications because of the high rate capability due to the lower resistance, but the amount of electrolyte held by the separator is subsequently lessened, which may increase cell resistance locally if the electrolyte is not uniformly distributed over the electrode surface.

There are some more fundamental properties to consider in the separator design process. Porosity is one of the essential properties that enable the separator to hold a large amount of electrolyte and not affect conductivity negatively. Nevertheless, a compromise between porosity and mechanical strength is often needed. Furthermore, the pore size and distribution should be uniform to ensure uniform current distribution over the entire electrode surface, and structurally the separator should have sufficient porosity to absorb the liquid electrolyte, and wet easily. The permeability of the separator is crucial, and should be as regular as possible to ensure uniform current distribution and uniform performance throughout cell life. The permeability is dependent on the porosity of the material, and can be seen as the effective capillary length of the separator.

The macroscopic structure of the separator should remain and not crumple even if soaked with electrolyte, and it must also be crease proof to keep the interface between the electrodes and the separator intact. These properties should be valid over a wide temperature range, but as the temperature rises, sooner or later the separator will reach its softening temperature at which point it tends to shrink, even for low porosity separators.

All separators will, however, degrade during cycling and storage. The primary cause is the degradation associated with the clogging of the pores caused by degradation products from the electrolyte (mainly SEI related) blocking the ion transport during cycling, and increasing the cell impedance. The performance of the separator can be improved by modifying the surface of the separator by, for example, wetting surfactants or hydrophilic functional components. Polymer coatings on the surface can also be applied to further improve the interface between the electrodes and the separator.

Another function of the separator is to act as an internal passive safety protection device. If overheated, by internal or external factors, the separator has the ability to shut down the cell from the inside, functioning as an internal thermal fuse of the cell. This shut-down property of the separator can also be triggered in overcharging conditions. Near the melting point of the separator, the pores collapse accompanied by a significant increase of cell impedance, resulting in a termination of the electrochemical reactions

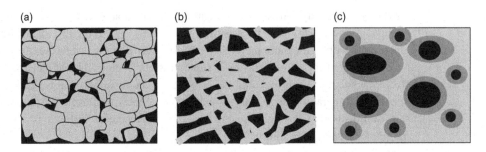

Figure 3.27 Schematic structures of the different types of separator materials: (a) inorganic composite membranes; (b) non-woven fabric mats; (c) microporous polymer membranes.

and of the current flow. It is possible to tailor the separator material used in order to optimise the shut-down response, and efficiency. Even if the separator melts, it should still be mechanically robust above the shut-down temperature in order to avoid an internal short circuit.

Another version of the shut-down mechanism is the use of a multi-layered separator. In such materials, the inner layer melts and fills the separator pores to inhibit ion transport, while the outer layers (i.e. near the electrodes) remain mechanically unaffected until their melting temperatures are reached. It is also possible to mix different polymeric compounds in a single-layer separator in which some areas melt and some areas just increase the cell impedance. It should be noted that by this shut-down mechanism, the cell is irreversibly damaged and cannot be used further. The shut-down separator is therefore the safety device inside the cell acting as a 'last line of defence'.

There are mainly three types of separator materials of interest for Li-ion cells employing liquid electrolytes: (i) *inorganic composite membranes*, (ii) *non-woven fabric mats*, and (iii) *microporous polymer membranes*. Figure 3.27 illustrates the schematic differences between the three types. There are several advantages and disadvantages of the different types, and it is also possible to make combined separators, but the drawback might be the extra space required, resulting in unnecessarily large cells.

The *inorganic composite membranes* exhibit high thermal stability and are formed by highly hydrophilic and wettable inorganic particles to form membranes of desired properties of thermal and dimensional stability. Often alumina, silica, or zirconia, or a mixture thereof is used. The drawback of this kind of separator is the brittleness at thin dimensions, and the lacking mechanical integrity during cell production.

A separator of high porosity is achieved by the *non-woven fabric mat* made of synthetic or natural fibres. Such mats exhibit larger pore sizes in a thin structure, and therefore lack the mechanical stability required for liquid electrolytes if not made in considerably thicker dimensions.

Microporous polymeric membranes are the separators mostly used in Li-ion cells. These membranes are mainly based on polyolefin compounds, such as polyethylene, polypropylene, or mixtures thereof. Such separators exhibit high chemical and mechanical stability, and are often made of a single layer or by several layers where each layer has a specific role in the overall cell performance.

3.3.3   Polymer-based electrolytes

Polymer-based electrolytes in Li-cells are of either a solid or gelled character: *solid polymer* or *gel polymer electrolytes*. These electrolytes also serve as separators; no individual separator is required. A solid polymer electrolyte only consists of a Li-salt dissolved in a polymer matrix. A gel polymer electrolyte, on the other hand, consists of a liquid solvent added to the polymer, in order to facilitate the ion conductivity, and thus the amount of polymer is lower and is mainly included for mechanical purposes. Cells made of either of these electrolytes are leakage free and may be easier to assemble than cells utilising liquid electrolytes. The absence of a separator can result in improved energy and power densities due to a lower mass of inactive and supportive materials.

The polymers used for *solid polymer electrolytes* are mainly ether-based polymers such as poly(ethylene oxide) (PEO) or poly(propylene oxide) (PPO). These electrolytes can be made thin and of flexible sizes and possess high safety due to the absence of flammable organic solvents. The ion conductivity relies on the amorphous character of the polymer and there are two main theories regarding the behaviour of the ion conductivity: Li^+ hopping from site to site or Li^+ transport associated to motions in the polymer due to local structural relaxations. The latter will, however, affect the Li^+ and the anions.

The ion conductivity of a solid polymer electrolyte is acceptable for temperatures above 60 °C, and therefore cells based on this type of electrolyte are preferably used in high-temperature Li-metal cells. Increased operational temperatures can be an advantage in the usage of electric vehicle applications and might result in simplified overall thermal management of the vehicle and simplified packaging.

When using solid polymer electrolytes in cells operating at lower temperatures, a wider temperature range of the amorphous phase is desired, and therefore a polymer of low glass-transition temperature is required, which could be achieved by different additives or mixed polymers. It is possible to incorporate inert inorganic nano-sized particles as additives, *plasticisers*, and thereby extend the amorphous region. These plasticisers will prevent the polymer chain from rearranging and crystallising, and they enhance the ion transport and the mechanical strength of the electrolyte. They may also dissociate the salt to a higher degree and suppress the polymer chain motion, which will decrease the anion transport and thereby increase the Li^+ transference number. The particles can, however, act as catalysts for electrolyte decomposition and should be added with care. Examples of this type of additive are Al_2O_3, TiO_2, SiO_2.

The lack of liquid solvents in the solid polymer electrolytes gives rise to different interface properties towards the electrodes, affecting the ion conductivity and performance rate of the cell. In a liquid electrolyte, the porous electrode is well wetted by the electrolyte and the ion transport across the interface can proceed smoothly and fast. Using a solid polymer electrolyte, it is difficult to fill all the pores in the composite electrodes, and thereby ion transfer is severely limited, resulting in large impedance increase and capacity loss due to the reduced contact area between the electrodes and the electrolyte. This limits the usage of solid polymer electrolytes in cells of highly porous electrode surfaces, like most composite electrodes used in Li-ion cells, and the solid polymer electrolyte is preferably used in combinations with a Li-metal negative electrode. Other challenges of solid polymer electrolytes are the electrochemical stability towards positive electrodes of high potentials (>4 V vs. Li/Li$^+$) and the manufacturing of uniform and defect-free large-format polymers of a thickness in the μm scale.

The *gel polymer electrolyte* is a ternary concept of Li-salt, liquid organic solvent, and polymer, where the salt and the solvent are embedded in a polymer network. These electrolytes have very much in common with the liquid electrolytes in terms of the ion conductivity and electrochemical stability. The polymers in a gel polymer electrolyte are more or less of an additive character, serving only as the mechanical matrix. The amount of polymer affects the mechanical strength, and chemical or physical cross-linking is often used. Due to the presence of liquid solvent, the interfacial behaviour is similar to liquid electrolytes in terms of, for example, ion transfer, formation of SEI layers.

Other types of polymers than those for solid polymer electrolytes, and often more stable against the electrodes, can be used. Typical polymers are poly(acrylonitrile) (PAN), poly(methyl metaactylate) (PMMA), poly(vinylidene fluoride) (PVdF), or similar compounds. The usage of PVdF may be beneficial to the ion conductivity due to the fact that the binder of most composite electrodes is PVdF, and it will consequently improve the interfacial properties.

In commercial context, Li-ion cells with gel polymer electrolytes are often named 'polymer Li-ion cells' or 'Li-ion polymer cells', which can be misleading by implying that a solid polymer electrolyte is used.

3.3.4   Ionic liquids as electrolytes

An ionic liquid consists of ions only and is a salt in its liquid state, i.e. a salt of low melting point. It is also a solvent. A general definition of an ionic liquid is that the melting point is below 100 °C, and regarding electrochemical cells the ionic liquid is preferably liquid at room temperature or below. The often high thermal and electrochemical stabilities are attractive characteristics for electrolytes for Li-ion cells. Due to the ionic character of the liquid, low vapour pressure is achieved, resulting in a high boiling point and a non-flammable liquid. This wide liquid temperature range and high safety level is very attractive for electrolytes of cells intended for electric vehicles.

Salts of a low melting point commonly consist of large bulky and asymmetric ions having a high degree of charge delocalisation. In ionic liquids, it is the size of the ions that decreases the ion–ion interactions, preventing crystallisation and keeping them liquid. The cations are often organic and the anions inorganic and the properties of the ionic liquid are strongly dependent on the cation and anion used. By combining cations and anions, ionic liquids can be tailored to suit different application requirements, e.g. compatiblity with metallic lithium electrodes.

To be of interest as the electrolyte of electrochemical cells, an ionic liquid should be liquid over a wide temperature range and be able to dissolve Li-salts, the latter as there is no ionic liquid with Li$^+$ as the cation. The viscosity of an ionic liquid is quite high, compared to liquid electrolytes, and adding a Li-salt results in even higher viscosities, often too high for the ion conductivity preferred in the cells. Therefore, in practice the ionic liquid-based electrolyte must be mixed with other types of solvents, often the same organic solvents as used in conventional liquid electrolytes, in order to obtain viscosity enabling acceptable ion conductivity. Ionic liquids can thus be seen as something in between an additive and a co-solvent, and the ratio between the organic solvent and the ionic liquid is a trade-off between the ion conductivity and safety (i.e. the non-flammability character). The salts used in ionic liquid-based electrolytes are similar to those employed in liquid electrolytes based on organic solvents.

4   Cell design

Cell design is largely the task of combining active and inactive cell components to make a complete cell usable in electric vehicle applications, something that requires an extensive optimisation process. Since there is no single combination suitable for all vehicle applications, usage behaviours, or performance requirements, different combinations of materials are possible. In this chapter, the cell design constraints will be discussed in terms of active materials, electrode design, and how to make the complete cell ready to be incorporated into a battery. Initially, focus will be on the design of the composite electrodes, which is essential for how the cell can be optimised towards energy or power. The electrode design and manufacturing are also the basis for durability and safety.

All-electric vehicles are, however, dependent on the power and energy performance of the cells but, as previously described (Section 1.6.4), these two performance characteristics cannot be combined without somehow oversizing the battery. A critical point in the cell development process is to match, or balance, the electrodes in order to obtain optimal cell performance during charge and discharge, exhibiting optimal capacity over a lifetime. Furthermore, size and format of the cell may be significant with regard to the performance of the cell, the battery, and subsequently the electric vehicle. Different cell formats will be discussed, as they also affect the thermal and packaging constraints of the final battery. Finally, a brief overview of the main production steps will also be given.

4.1   Composite electrodes

The active electrode material is mixed with an electron-conducting material and a binder to create a *composite electrode*. This is attached to a current collector, typically a thin metal foil. This basic set up of a composite electrode – negative or positive – is illustrated in Figure 4.1, displaying the electron conduction pathways, where the electrons are conducted from the active material, via the conductive material, to the current collectors. The ions are conducted via the electrolyte, which is incorporated in the voids between the particles, to the active electrode materials.

The size and distribution, surface area, and morphology of the particles of the active materials are all essential parameters for optimisation, and can enhance the redox reactions and thus improve the energy and power performance, as well as the stability

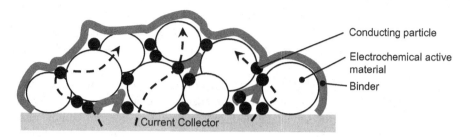

Figure 4.1 Illustration of a composite electrode, indicating electron conduction pathways in the electrode and the active electrode material.

and durability. Moreover, the porosity of the final composite electrode is significant in enabling the electrolyte to be incorporated into the electrode and further to enhance the ion conductivity and minimise the interfacial impedance. The porosity is often in the order of 30%.

Nano-particles will shorten the diffusion lengths and increase grain boundary diffusion. Shorter lithium diffusion lengths in these particles enable acceptable rate capabilities in materials exhibiting limited performance in their micro-sized states, such as $LiFePO_4$. The nano-sized materials should, however, be handled with care since the increased surface area of the particles, compared to larger sized particles, could catalyse side reactions and decomposition of the electrolyte, which may result in severe safety concerns. Thin coatings on the particle surfaces can improve electronic conductivity, decrease reaction rates between the electrode and the electrolyte, stabilise the active material towards dissolution, suppress loss of oxygen, or increase the potential. The challenge is to find coatings that prevent reactions with the electrolyte, yet are still able to transport lithium ions at high rates. The coatings must also remain unaffected by the cycling conditions.

In order to enable insertion and extraction of lithium ions, proper contact between the particles of the active material within the composite electrode must be secured. Ion and electron conductivity between the particles is further increased by adding a conductive material, often amorphous carbon in the form of carbon black or whiskers of carbon. Alternatively, metal particles can be used. It is crucial that the added conductive material does not affect the electrochemical reactions or act as a lithium-trap, which will result in capacity losses. Nano-structuring and surface coatings are two commonly used methods allowing materials once rejected to actually function as active materials.

The *binder* material will further enhance the contact between the particles and hold the composite electrode and the current collector together. Polymeric materials, like poly-vinylidene fluoride (PVdF) or copolymers thereof, are often used to achieve a porous, but still dense, electrode. The binder should be very viscous and have a low glass transition temperature to ensure a high mechanical stability over a wide temperature range. It should also be chemically inert and stable in the electrochemical environment of the cell. As a non-active material in the cell, the amount of binder should preferably be as low as possible. Another feature of the binder is the elasticity needed to accompany the volume changes occurring within the active materials during the charge and discharge processes, without affecting the dimensions of the electrodes (e.g. the thickness). The dimensions are important to secure performance, i.e. exhibit the same charge-transfer properties at the electrolyte–electrode interfaces and between the active material particles.

To secure mechanical performance of the composite electrode the mixture of active material, conduction material, and binder will be attached to a *current collector*. The current collector should be designed to provide a uniform current distribution and exhibit low contact resistance to minimise electrode polarisation during cell usage. A crucial parameter for the current collector is compatibility with the active material, the binder, and the electrolyte without any degradation or corrosion. Furthermore, the current collector is a non-active cell component and should therefore be as thin and light as possible with high mechanical stability but without limiting the cell performance or the manufacturing process. All these requirements are not easily fulfilled for all operational conditions, and corrosion of the current collectors is one of the main factors responsible for a decrease in cell performance during the life of a cell. The consequences are mainly a decrease in the contact area towards the electrode and thereby an increase in the resistance between the composite electrode and the current collector. If severe corrosion products are formed, they may cause cracking of the electrode and consequently a power fade.

The potential of each electrode determines the current collector materials that are viable. Aluminium is preferred due to its high abundance and low weight, and is the current collector used in electrodes with a potential above *ca.* 0.5 V. At lower potentials, lithium will alloy with aluminium, causing a drastic decrease in the amount of cyclable lithium and thereby a decrease in cell capacity. Therefore, copper foil is used as the current collector of all carbon- and alloy-based negative electrodes. Negative electrodes based on $Li_4Ti_5O_{12}$ can, however, use aluminium due to the higher electrode potential. Moreover, the weight of a Cu current collector is significantly larger than that of an Al current collector, affecting the energy density of the Li-ion cell negatively.

The Al current collector of the positive electrode is constantly at potentials where the Al material is thermodynamically unstable. Therefore, a thin stability protecting film of oxide, oxyhydroxides, or hydroxides, is applied to the surface to passivate the Al-surface. This prevents corrosive pitting of the current collector during cell operation at potentials above *ca.* 3.6 V vs. Li/Li$^+$. Depending on the Li-salt used, the protective surface film on the Al current collector becomes unstable and corrosion processes will occur. Aluminium ions can be formed, dissolved, and transported to the negative electrode via the electrolyte, and may impair the performance of the negative electrode by blocking the lithium diffusion pathways or interfere with the SEI layer.

During the production processes, of active materials and composite electrodes, impurities may be incorporated. These unwanted compounds can be in the form of particles or, for example, moisture, which may interfere with the redox reactions or cause decomposition reactions. For example, the surface of the Cu current collector will lose stability if traces of HF are present in the electrolyte and acid-base reactions take place between the protective oxide layer and the HF impurities. Various protective adhesion improvement layers can be used on the current collectors, but like all non-

(a) (b)

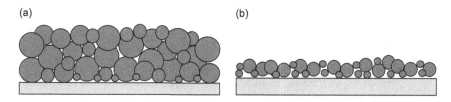

Figure 4.2 Schematic picture of (a) an energy-optimised electrode and (b) a power-optimised electrode.

active materials added to the cell they should be compatible and stable against the electrochemical environment in the cell.

4.2 Energy and power-optimised electrodes

The energy and power performance are the most important parameters of a cell for vehicle applications, but they cannot be combined in a single cell since the cell is optimised for either energy *or* power. From a vehicle perspective, power corresponds to acceleration and energy to the all-electric driving range. The energy content in a cell relates to the capacity of the material used, and the power capability is mainly related to physical aspects of the active materials and the electrolyte design. It is possible to calculate the theoretical energy density based on cell voltage and theoretical capacity (equation (1.28)), whereas it is impossible to calculate the theoretical power density since this is not a fundamental material property, but related to many different parameters interacting: e,g, particle size, morphology, surface area.

How the composite electrodes are fabricated is thus crucial in terms of energy and power density of the cell, and there are some general constraints to be considered in the design of the electrodes in order to optimise the cell. The energy and power density largely depend on the porous microstructure of the composite electrode: size, shape, and morphology of the active material particles, the binder, and the preparation methods. The *tap density*, approximately the inverse of the porosity, indicates the energy and power performance and is a control parameter in cell manufacturing; electrodes having high tap density will provide high energy, and vice versa. In Figure 4.2, schematic representations of an energy and a power-optimised electrode are shown.

The energy density of the cell is determined by the capacity loading of the cell, i.e. the amount of active material, and thus thicker and denser electrodes are required for increased energy density. In this optimisation process, the weight of the different components needs to be considered, especially the weight of the non-active components such as current collector or binder. Generally, the energy density of a cell is the reversible charge transfer between the electrodes as a function of voltage. The charge transfer decreases with increased C-rates. At high C-rates, the ion diffusion within the electrodes and across the electrode/electrolyte interfaces is slower than the charge distribution to reach equilibrium, i.e. the materials are not fully utilised and cause a decrease in the reversible cell capacity. Thus, as a result of optimising the cell for higher energy densities, a low rate capability is achieved.

In order to enhance the power capability of a cell, voltage drops caused by an increased current should be minimised: ohmic, kinetics, and mass-transfer limitations need to be overcome, as illustrated in Figure 1.23. The internal cell resistance and impedance should be reduced, and there are several routes; e.g. cell impedance is complex and can vary with time, current (rate and direction), voltage, temperature, etc. Also, particle size plays an important role for cycling stability, capacity, and the power capability of the cell; some materials exhibit low electronic and ionic conductivity in the micro-sized state, but perform well in nano-sized state. Nano-sized particles are, however, not always attractive, since they may act as catalysts of electrolyte degradation, especially in combinations with large polarisations. Furthermore, the electrode kinetics is mainly limited by the lithium ion diffusion within the porous electrodes. Therefore, the power density and the rate capability of an electrode increase with decreasing electrode thickness (i.e. the energy density will be reduced accordingly). A thin electrode often requires a thicker current collector to obtain the same discharge capacity, which results in an increased weight of inactive material and consequently a decrease in energy and power densities.

4.3   Energy and power-optimised cells

After selecting the combination of active electrode materials and optimising the electrodes towards energy or power, the next step is to assemble the cell. The cell can be tailored to fit specific performance requirements, and some material combinations are more suitable for power and some for energy, all depending on the kinetics. Besides the material combination, i.e. the active materials of negative and positive electrodes, the matching of the electrodes, *cell balancing*, is key to producing a cell of optimal usage characteristics.

4.3.1   Cell balancing

Once assembled, the Li-ion cell is in a fully discharged state and all cyclable lithium ions are found in the positive electrode. During the first cycle, the *formation stage* of the cell, some lithium is consumed and an irreversible capacity loss is a fact. Depending on the active materials in the electrodes, different lithium-consuming processes take place. The most common is the SEI formation on the carbon-based negative electrode. Therefore, the negative electrode is generally larger (in capacity and size) than the positive electrode to compensate for the irreversible capacity loss resulting from the SEI formation. The reversible capacity of the cell is consequently determined by the amount of accessible lithium in the positive electrode material, and the capacity of the negative electrode is never fully utilised. Typically the negative electrode is 10–20% larger; power-optimised cells in the upper range and energy-optimised cells in the lower range. Furthermore, if the negative electrode cannot compensate for the irreversible capacity losses, lithium plating may take place, affecting the safety of the Li-ion cell.

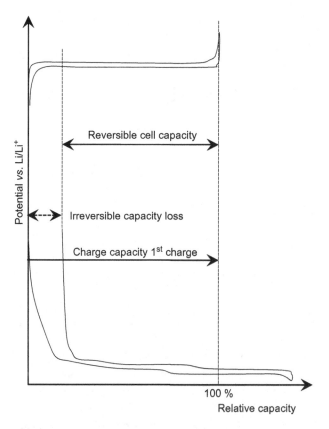

Figure 4.3 The cell capacity decrease during the first charge and the usable cell capacity of a Li-ion cell of graphite|LiFePO$_4$.

The irreversible capacity loss and the corresponding cell capacity is illustrated in Figure 4.3. A Li-ion cell of graphite|LiFePO$_4$ will serve as an example. During the first charging process of the cell, the cell reaches a fully charged state as all lithium ions have been extracted from the positive electrode. Most of the lithium ions will be inserted into the graphite structure, while a minor portion will take part in the SEI formation. The subsequent discharging process will use the remaining amount of cyclable lithium ions, and the first discharge capacity will therefore be lower than the first charge capacity. Thereafter, in an ideal cell, the usable cell capacity will remain more or less constant. The cell balancing will also limit the durability and safety of the cell. Durability will be affected by the SEI growth during cycling and will cause the cell balance to shift with time and usage.

4.3.2   Energy and power relationship

Cells optimised for either power or energy have different characteristics, characteristics difficult to combine in one cell. The usable capacity is affected to a considerable extent by the C-rate (Figure 1.26), and, for example, the shapes of the curves in the Ragone

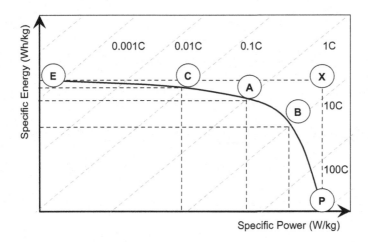

Figure 4.4 Schematic Ragone plot of an energy-optimised and a power-optimised cell, respectively. Different C-rates are indicated: A) 1C, B) 10C, C) 0.1C. X refers to an unachievable state. Maximum energy and power densities are marked by E and P.

plot (Figure 1.28) reflect the power and energy ability of the cell. Usually, the maximum energy and power densities of a cell are given (E and P, respectively, in Figure 4.4), or at a specific C-rate, e.g. 1C (A in Figure 4.4). The C-rate will affect the energy and power output. For example, the corresponding usable energy will be drastically reduced as the C-rate increases (B in Figure 4.4). On the other hand, high energy can only be utilised at low C-rates (C in Figure 4.4), and maximum power can only be utilised with infinitesimal energy, and vice versa. Consequently, coexistence of maximum power *and* energy (X in Figure 4.4) is fundamentally unachievable. Figure 4.4 illustrates how the capacity can be utilised if the cell is optimised towards energy or power.

4.3.3   Example: energy and power-optimised cells

In order to illustrate the differences in the cell properties of energy and power-optimised Li-ion cells, a few examples will be given. By keeping all cell parameters constant, except for the active material of the positive electrode, relevant differences in cell weight and energy density can be calculated. The negative and the positive electrodes are both made of 90% active material, 5% carbon black, and 5% binder, and have a porosity of 30%. The current collectors used are 20 μm thick Al-foil for the positive electrode and 15 μm thick Cu-foil for the negative electrode. The N/P ratio is defined to 1.2, and all other cell constraints are fixed, i.e. separator, electrolyte, and casing, as well as the physical size and format. The usable capacity of the negative graphite electrode is the same in all cases (330 mAh/g).

Three types of active materials of different capacity (LMO, NMC, and LFP) are used as the positive electrode. Two cells of different capacity are used: a power-optimised cell of 5 Ah and an energy-optimised cell of 25 Ah. The results are summarised in

Table 4.1 Positive electrode and cell characteristics for a power-optimised cell of 5 Ah for HEV applications utilising different active materials.

	LMO	NMC	LFP
Capacity of material (mAh/g)	100	150	150
Weight of electrode (g)	55	37	37
Weight of cell (g)	175	200	140
OCV at 50% SOC (V)	4.0	3.7	3.2
Energy density of cell (Wh/kg)	114	93	114

Table 4.2 Positive electrode and cell characteristics for an energy-optimised cell of 25 Ah for EV applications utilising different active materials.

	LMO	NMC	LFP
Capacity of material (mAh/g)	100	150	150
Weight of electrode (g)	280	185	185
Weight of cell (g)	630	540	580
OCV at 50% SOC (V)	4.0	3.7	3.2
Energy density of cell (Wh/kg)	159	171	138

Tables 4.1 and 4.2, respectively. Please note that these data are estimated using a tool designed for the optimisation of the battery characteristics of different electric vehicle types.[1] The calculations are based on material data and the presented cell data should only be seen as illustrative examples of how different material constraints affect final cell characteristics.

As can be seen from the data in Table 4.1, the differences in material capacity will not be directly reflected in the resulting cell weight and energy density. Comparing the LMO and LFP materials in a 5 Ah cell, the same energy density is achieved, even if the capacity of the materials differs by 50% (!).

On the other hand, energy-optimised cells of 25 Ah based on the same positive electrodes as in the 5 Ah case will show other relationships, as given in Table 4.2. Here NMC is clearly more suitable for energy-optimised cells. LFP has the same material capacity as NMC, but the LFP cell has an energy density even lower than the LMO cell, which has 2/3 of the capacity at the material level. This implies that the *material* with the highest capacity will not necessarily result in the *cell* with the highest energy density, and the full cell has to be considered in any comparisons.

If a cell is optimised for power or energy, this can affect the energy density differently depending on the active material used. Moreover, the size of the cell (in Ah) also affects the energy density. In the present example, the 25 Ah cells have higher energy density compared to the 5 Ah cells; the weight of the inactive materials has to be considered. Therefore, it is important to understand how the electrode materials will affect cell

---

[1] The BatCaP calculation tool, developed by Argonne National Lab, USA.

performance and how to select the most optimal cell for a specific task. In Section 5.2.2.2, some of these cells will be used to illustrate how cell selection will affect battery performance, and how non-optimal cell selection can drastically reduce battery performance, depending on the type of electric vehicle.

4.4   Cell format and design

There are a number of cell formats, ranging from small consumer cells, used, for example, in hearing aids, to very large cells for industrial back-up systems. Besides the composite electrodes and the electrolyte, the cell consists of several other parts, and is encapsulated in some kind of casing to hermetically seal the cell. The casing material is usually based on aluminium or plastics, or a mixture thereof and due to the sealed construction a Li-ion cell does not require any maintenance.

Cell formats and sizes suitable for electric vehicle applications can vary depending on application and usage condition, as well as manufacturer. The geometric formats of the cells suitable for electric vehicles can, however, be grouped into the following classes: *cylindrical*, *prismatic*, and *pouch*, as illustrated in Figure 4.5. These formats have different performance characteristics and constraints both from the battery and vehicle perspectives.

4.4.1   Cylindrical cells

Cylindrical cells are wound like a jellyroll, and the diameter of the cylinder indicates the capacity: thicker cells mean more capacity, and vice versa. The nomenclature used is often based on the diameter and the height of the cylinder and common cell formats are the 18650 cell (diameter = 18 mm, height = 65 mm) and the 26650 cell (diameter = 26 mm, height = 65 mm). An aluminium casing is often used and the terminals are found at the top and the bottom of the cylinder.

The simple production of cylindrical cells is one of the main advantages of this format, while the temperature management and packaging issues are the main

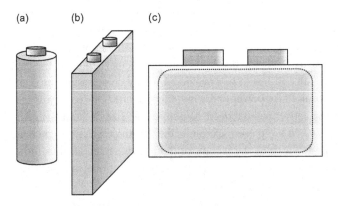

(a)   (b)   (c)

Figure 4.5 Different cell formats: (a) cylindrical, (b) prismatic, (c) pouch.

drawbacks. During cell usage, the internal cell temperature will increase due to Joule heating (Section 1.4.3). The temperature within the cell is not uniform; the core will be warmer than the shell, depending on the ambient temperature. Since optimal performance requires a uniform temperature distribution, heat needs to be removed from the core in order to avoid abuse situations, such as a thermal runaway. The size of the cells is therefore a trade-off between capacity and thermal performance, and consequently thinner cells are often used in power-optimised cells.

From a battery perspective, cylindrical cells cannot be packed in a space-optimal way when they are connected in series and/or parallel into a box intended for vehicle installation. There will always be voids between the cells, and, depending on the performance parameters, the final size of the battery may be unreasonably large. The voids are, however, desired in terms of thermal management of the battery and can, for example, accommodate cooling channels. Moreover, the position of the terminals may also affect the packaging of the cell: the terminals are found on the top and the bottom of the cylinder, which should be compared to prismatic cells (see below).

4.4.2 Prismatic cells

Prismatic cells can be produced in two ways: wound or stacked, depending on the performance characteristics required. Z-folded prismatic cells can also be used. The casing is often made of aluminium or reinforced plastics, or a soft casing: the pouch cell (see below). There are a number of cell sizes, and the nomenclature is not as standardised as that of cylindrical cells and may differ widely among manufacturers. Normally, however, the format is given in terms of thickness, width, and length, and there is a trend towards thinner and wider cells in order to facilitate thermal management.

From a packaging perspective, prismatic cells are attractive, and batteries of high density can be achieved. The packaging density can, however, result in abuse situations if the thermal management is not properly designed and controlled. Even if the temperature differences within the cells are more uniform in these cells, the heat dissipation from the cells can be an issue, and adjacent cells may heat each other. Therefore, the prismatic cells are arranged together with thermal management devices in between (some of) the cells. Furthermore, the casing surface of a prismatic cell can be designed to incorporate, for example, air flow channels, thereby combining the packaging density and the thermal management. Moreover, packaging efficiency is enhanced if the electric terminals are found on the same side of the prism, as is also indicated in Figure 4.5b.

4.4.3 Pouch cells

Pouch cells, often referred to as coffee-bag[2] cells, are prismatic or flat in the basic format and encapsulated in a soft package casing, often a polymer-laminated aluminium foil. Inside the cell, electrodes are mainly stacked or Z-folded to achieve the desired cell

[2] Coffee-bag package materials was used in the early days for research cells at Uppsala University, and hence this name.

capacity and energy and power characteristics. The flexible cell package is sealed by heat or ultrasonic welding. Like the hard-casing prismatic cells, the temperature differences within the cells are small, but thermal management between the cells is still required for safe operation. In addition, the welds may be of concern regarding robustness and durability.

The *tabs*, i.e. the terminals of pouch cells, pass through the seals at the edges, and can be located on the same or opposite side, as illustrated in Figure 4.6. Depending on their position, different cell performance characteristics can be obtained. If the tabs are placed on opposite sides, preferably on the short sides, the cells are often utilised as power-optimised cells, and if the tabs are on the same side, the cells are mainly utilised as energy-optimised cells. The position will also affect the packaging possibilities and complexity. If the tabs are on the same side, connecting devices are required on this one side only and thereby the electronics can be simplified and concentrated in the battery design, making the battery more condensed and easier to handle.

In Table 4.3, a comparison of the different cell formats and their properties are given. It should be noted that this is a generalisation and may differ among manufacturers and by the chemistries used.

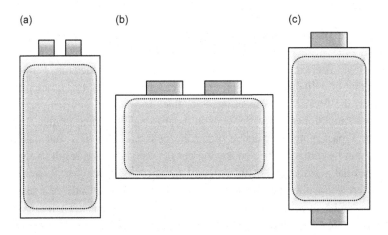

Figure 4.6 Possible positions of the tabs in a pouch cell; (a) on the same short side, (b) on the same long side, (c) on opposite short sides.

Table 4.3 Overall performance comparison of Li-ion cell formats

	Cylindrical	Prismatic	Pouch
Packaging	−	+	0
Gravimetric energy and power density	0	0	+
Volumetric energy and power density	0	0	+
Cooling options	−	+	0
Safety	+	+	0
Robustness	+	+	0

4.4.4   Cell safety devices

The cell can also be equipped with some, more or less sophisticated, safety devices, related to pressure release and/or current interruption, e.g. safety vents or overpressure devices. Two commonly used devices are the *current interrupt device* (CID) and the *positive temperature coefficient* (PTC) contact discs.

The CID is incorporated into the cell to interrupt the electrical connection between the positive electrode tab and the corresponding terminal when the internal cell pressure, due to gases produced by an abuse situation, reaches a given level. Another common safety feature of Li-ion cells is the cell vent, which is activated at pressures just above the CID-level to release the gases and stop the charging/discharging process of the cell.

The PTC, on the other hand, is a device that increases its resistance at a given temperature or current flow. At normal currents and temperatures, the PTC has a very low resistance. When excessive current passes through the device or the cell temperature increases rapidly, the resistance increases by an order of magnitude, drastically limiting the current. These devices are incorporated in some of the cells in a battery or at other positions along the current distribution network within the battery. A PTC device will, however, not prevent a short circuit condition from discharging the cell or the battery fully.

4.5   Production processes

The energy density of Li-ion cells has increased over the years, often due to progress in cell construction and production processes rather than to improvements in the materials. Progress in production techniques has also been significant for increasing the robustness and safety of the cells. In the case of EVs, it is important to realise that constructing larger cells is far from simply up-scaling smaller cells for portable applications; other requirements in terms of quality and performance come into play.

The production process is similar among manufacturers, and in the following section the main production steps and their relation to cell performance will be described. A brief overview of the production process is given in Figure 4.7. It covers the main steps from raw material handling, via mixing, coating, calendaring, winding/stacking, filling of electrolyte, sealing, formation, sorting, ageing, and shipping. All steps are required to be closely monitored for performance, safety, and durability of the final product. The processes can be divided into three main parts: *electrode fabrication, cell assembly*, and *cell formation*.

The *electrode fabrication* is based on coating processes, where the active materials are coated onto the current collector. The active material is formed by mixing the conductive additive and the binder material together with a solvent into a highly viscous slurry. Several factors affect the electrode design and the overall performance properties of the Li-ion cell, and the particles of the active material are key: size and distribution, shape, morphology, and specific surface area, as well as the tap density of the electrode.

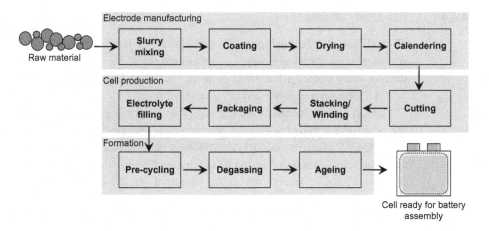

Figure 4.7 Rough overview of the cell manufacturing process.

Therefore, the synthesis of the active material is one crucial step in the pre-cell production sequence, and very small differences between batches and manufacturers can result in large cell performance variations.

The adhesion of the slurry layer to the current collector is crucial for low resistance, and the coating can be performed according to several techniques, depending on the final cell properties, and varies among manufacturers. The common requirement is the ability to produce homogeneous coatings, in terms of both the thickness and tap density. Usually, the slurry is coated on both sides of the current collector, with a uniform distribution of the active mass. The thickness of the coating is normally within the range of 50–300 μm, the exact value being dependent on whether the cell is optimised for energy or power.

During the drying process, evaporation of the solvent of the slurry, crack formation, and peeling of the electrode layer from the current collector are risks that should be avoided. If water is used as the solvent, it is important to ensure that no traces of moisture are left, as these may cause severe damage and fast degradation of the cell. After the drying process, the electrodes are calendered for the right thickness and porosity by being compressed to the appropriate thickness. In some cases, a thin protective layer of, for example, Al_2O_3 is coated on the electrode surface to increase the stability and reduce the likelihood of internal short circuit.

Depending on the intended format of the final cell, i.e. cylindrical, prismatic, or pouch, the *assembly process* can vary, but the composite electrodes must withstand the mechanical constraints during the assembly process. An energy-optimised cell has a denser electrode, and the thickness can be limited by the manufacturing process; cracking and flake-off during the cell assembling process must be avoided. The stacking, or winding, of the electrodes often involves several steps where the separator and the electrodes are sandwiched into the final shape of the cell and packed into the casing.

As a final step before the cells are sealed, the electrolyte is filled. This is a critical part in the cell assembly process, and is a trade-off between minimising the free amount of electrolyte and optimising the electrode/electrolyte interface. The electrolyte filling step and amount of electrolyte for a given cell varies among manufacturers.

Once assembled, the cells are passed on to the *cell formation* stage. As the Li-ion cell is discharged, it must be activated by charging. The first charge is called formation and activates the active electrode materials, often at slow charging rates (i.e. using low currents) to properly form the SEI layer on the negative electrode. During the formation step, electrolyte will diffuse over the entire electrode surface to fill all pores and produce an optimal interface between the electrodes and the electrolyte (or the separator). A proper wetting of the electrodes and the separator is desirable for low interfacial impedance of the cell.

The formation process may in some cases last a few full cycles, depending on the manufacturer and the purpose of the cells. Fully charged, the cells are stored, or aged, for a predefined time period at a specific temperature set by the manufacturer. After storage, the cells should have cell voltages that are as uniform as possible. Cells may be short circuited due to fabrication malfunctions, like impurities in the slurry. This will be detected as voltages that are too low after the ageing. Therefore, all cells should be checked after storage in terms of quality and performance, including safety.

4.5.1   Safety and reliability

The design and manufacturing may influence the safety of the final cell. Safety and reliability are related to the materials, the design, and the usage of the cells. Regarding the materials used, the interfacial interactions between the electrolyte and the electrodes, the temperature, and the state of charge of the cell are significant factors. The selection of separator material is another aspect of stability and shutdown performance. From design and manufacturing perspectives, different steps can be taken in order to improve safety and reliability; electrode balancing, coating uniformity, and geometrical design of the electrodes. Moreover, the position and insulation of the tabs, and the terminals are essential to, for example, prevent short circuits.

During the manufacturing process, there are additional factors to consider in terms of safety. The selection of raw materials is the first one; the materials should be as pure as possible. Maybe one of the most important factors of the manufacturing processes is the cleanliness of the manufacturing facilities in order to minimise the risk of impurities being incorporated into the cells, including all the handling in between different processes. The impurities may, depending on their character, composition, and shape, act as short circuit germs to facilitate side reactions. A transparent manufacturing process is desirable in order to secure cells of high reliability and quality. The production of cells aimed for electric vehicles should ensure that no more than $1/10^8$ is a failing cell, which is one or two orders of magnitude lower than in the case of cells for consumer electronics.

III

Battery usage in electric vehicles

In the previous two parts of this book, electrochemical and material properties of the battery cells have been covered, focusing on Li-ion cells. Li-ion cells and their properties will now be the basis for the remainder of the book focusing on battery design and usage in electric vehicles, where fundamentals and material constraints also are turned into observations on the battery and vehicle levels.

In the following chapters, the vehicle requirements and usage conditions affecting the battery are the focus. To begin with, vehicle requirements and processes at vehicle level are discussed. Thereafter battery design, including cell selection, and battery management will be described from a cell perspective. Finally, cell degradation is the main topic and general as well as specific degradation mechanisms concerning cells – composite electrodes, active electrode materials, and electrolytes – will be described in order to illustrate the complexity of battery usage. Li-ion cells will be the technology used here in order to illustrate usage, performance, and degradation of batteries for electric vehicles.

5 Vehicle requirements and battery design

Different vehicles have different needs and the corresponding requirements of the battery vary accordingly. No battery can be made that is able to fulfil all the functional and operational requirements of all types of electric vehicles. From a vehicle perspective, the battery should obviously deliver and accept power according to the usage conditions, but the battery should also be robust and reliable, possess long durability, and require minimal service and maintenance.

A battery is also more than just the cells contained. In order to be functional in a vehicle, the battery must include a thermal system, electronics, and a management unit. Vehicle performance will strongly affect the battery design and the selection of battery technology; design constraints that influence reliability and driveability. Consequently, exactly the right cell and battery can only be chosen with the electric vehicle in mind, to avoid oversized batteries – unnecessarily heavy and expensive. Examples will be given on how to select battery and cells in order to fulfil the vehicle requirements.

5.1 Vehicle types and requirements

The primary advantages of electric vehicles are environmental: reduced fossil fuel consumption, and hence CO_2 emissions, and reduced pollution and noise levels in cities. Moreover, vehicle integration may be simplified since no mechanical connections are needed between the electric drive system and the battery. Another advantage is the ability to recuperate brake energy, which is free energy to be used in the overall energy management strategies of the vehicle.

The basis for all battery development for electric vehicles is the vehicle type and how the vehicle is to be used. Depending on how the electric vehicle is designed, different battery requirements are needed in terms of power and energy needs. These requirements will be discussed in the following sections, starting with the main types of electric vehicles.

5.1.1 Vehicle types

Electric vehicles can have different degrees of electrification, i.e. a certain amount of the energy consumption derives from electricity and the rest from the internal combustion engine (ICE), as illustrated in Figure 5.1. Several factors must be considered in order to

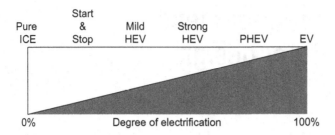

Figure 5.1 Degree of electrification for different types of vehicles.

select the optimal degree of electrification: required all-electric driving range, energy consumption per km, performance requirements (e.g. acceleration), geographical constraints such as ambient temperature and terrain.

Depending on the intended use of the vehicle, different degrees of electrification are used, and can vary among the same vehicle segment depending on the vehicle manufacturers, their business models, and the customers. As long as the degree of electrification is less than 100% an ICE, or a similar, is required. The battery is the sole source of energy only when the vehicle is 100% electrified.

The degree of electrification is a trade-off between installation complexity, performance, and cost, and somewhere along the scale there are breakpoints at which some solutions turn out to be more beneficial than others. These breakpoints depend on type of vehicle and manufacturer and the ability to design and utilise the battery in an optimal way for performance, safety, and durability. Different types of vehicles require different battery technologies capable of meeting specific vehicle constraints.

In order to simplify the battery requirements, different types of vehicles are grouped according to their degree of electrification. The battery design and the cell selection very much depend on the type of electric vehicle. Non-optimal choices may result in unnecessarily oversized batteries in order to fulfil the vehicle requirements.

5.1.1.1   All-electric vehicles

An *all-electric vehicle* (EV) is the simplest type of all-electric vehicles due to its single source of energy: the battery. Sometimes these battery-powered vehicles are called *battery electric vehicles* (BEV) or *pure electric vehicles* (PEV). PEV is a wider concept incorporating vehicles also utilising other electric energy sources, e.g. fuel cells. The battery and the electric drive system must single-handedly be able to fulfil all vehicle requirements of a conventional, non-electric vehicle; the electric motor must be able to accelerate and propel the vehicle, and the driving range is directly related to the capacity of the battery. The usage conditions, will, however, affect the driving range considerably; aggressive driving, air conditioning, and other energy demanding devices will consume energy and shorten the range, in the same manner as for a conventional vehicle.

The battery of an EV acts like the fuel tank of a conventional car; it can be filled up and the vehicle runs until the tank is empty. The drawback of the EV is the refuelling time, which is much longer. How 'full' the battery is ☐ referred to as the state of charge (SOC) ☐ is where a fully charged battery has a SOC of 100%, and an empty battery 0%. State of charge, and other state functions of the battery, will be described in more detail in Section 6.2.1. An EV is not utilising the full SOC range, but a wide SOC range is used and the range is a trade-off between battery durability and driving range. The energy is the performance parameter for an EV coupled to the driving range. The power demands are, however, to be considered for drivability, but also to enable fast-charging procedures. In order to facilitate the thermal management of the battery and the vehicle itself, a heating device is often installed in vehicles operating in cold climates. This assists in optimising battery performance and in avoiding severe degradation reactions related to low-temperature operation. These are both comfort and performance assuring devices that differ between manufacturers and regions.

5.1.1.2 Hybrid electric vehicles

A *hybrid electric vehicle* (HEV) is a vehicle having two or more types of power sources, at least one of them being electric. Brake energy is regenerated and stored in the battery, and is an important feature of an HEV in terms of total energy efficiency. The key factor of an HEV is the interactions between the different power sources; normally an ICE and an electric drive system. The battery in an HEV can only be charged by recuperating brake energy or utilising the ICE, i.e. no external charging procedure is required or even possible, making the HEV very much like a conventional vehicle in usage.

There are a wide range of HEV configurations with different degrees of electrification: from *micro* HEV to *mild* and *strong* HEV. In a micro-HEV, usually called 'Start-Stop', the ICE is normally automatically turned off whenever the vehicle is idling during coasting, braking, or stops, and then restarts quickly. When idle, the battery provides power to the auxiliary loads, e.g. the air-conditioning and radio systems. Often the voltage level of the battery in a micro-HEV is the same as for the auxiliary loads, i.e. 12 V or 24 V depending on vehicle type.

Hybrid electric vehicles with larger battery capacity can be categorised as either mild or strong HEVs. In a mild HEV, the battery provides power to the ICE operation and utilises regenerative braking, but no all-electric drive is possible. In a strong HEV, on the other hand, a limited all-electric driving range can be utilised with the distance depending on battery capacity.

Power is an important performance parameter both for mild and strong HEVs, during charge (recuperation of brake energy), as well as discharge (acceleration). The battery must be able to deliver and accept short power pulses of high amplitudes, from parts of seconds up to tens of seconds, and the power demands must be delivered/accepted within the specified voltage limits of the battery. The battery is usually of several hundreds of volts in order to maintain the currents at practical levels. For example, the diameter of a cable is related to the current carried, and for higher currents thicker cables are required, which eventually would result in cables difficult to handle during vehicle production.

5.1.1.3   Plug-in hybrid electric vehicles

The *plug-in hybrid electric vehicle* (PHEV) combines the advantages both of the EV and the HEV. The key features are the ability to charge the battery externally, a longer all-electric driving range than for a strong HEV, and an overall longer driving range than for an EV. Often a PHEV has three operational modes: pure ICE mode, pure electric mode, and hybrid electric mode. How these modes are utilised depends on the capacity and status of the battery. Therefore, a PHEV must be able to provide full acceleration power using solely electricity and the battery capacity is large compared to an HEV. Battery performance in terms of energy is smaller than an EV, but the power demands and possibilities are often in the same range as for an HEV.

The all-electric driving range depends on the capacity of the battery and can be a marketing feature to attract customers. The ability to recharge the battery externally by grid electricity, regenerative brake energy, or by the ICE are design features used to optimise energy consumption and thereby battery selection. These energy management strategies are key design features for HEVs and PHEVs.

5.1.2   Usage conditions

Before moving from the different types of electric vehicles to battery design, the usage conditions have to be considered. How the vehicle is operated under real driving conditions, i.e. the relationship between charging and discharging as a function of time, is important in battery selection. The conditions differ between markets, geographical regions, drivers, vehicle concepts, etc. In order to understand when and where EVs, HEVs, or PHEVs make sense and the optimal hybridisation degree, the usage conditions must be known or at least considered.

The usage conditions are ideally provided as concise sequences of vehicle operational data over a time period, typically second-by-second data of speed. These conditions can be obtained either from data logging of vehicles in real operation or they can be artificially designed according to standardised vehicle behaviours based on data from a large number of vehicles used under different operating conditions. Figure 5.2

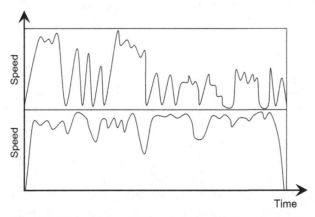

Figure 5.2 Urban (top) and highway usage conditions (bottom).

illustrates schematic usage conditions for urban and highway driving, respectively. From these data, the battery design parameters can be derived, e.g. maximum and average speed, idle time, frequency of starts and stops. The energy consumption per distance can be very different depending on the type of vehicle and the driver's behaviour, resulting in different battery demands.

On the other hand, a *duty cycle* represents how the vehicle will be operated in the target application, and is especially important for vehicles utilised most of the time (taxis, buses, etc.). The duty cycle data will account for more than just vehicle speed as a function of time. Cargo weight (e.g. number of passenger in a bus), road surface, route type, and vocational loads can also vary significantly over time. Moreover, other loads in the vehicle during operation, e.g. the air conditioning system, must be taken into account in the battery selection process.

In this context, the *energy throughput*, the total amount of energy delivered and accepted by the battery, often measured as the total energy per time interval (e.g. kWh per hour), is a useful measure, describing the average power needed in order to fulfil the duty cycle requirements. Normally, the battery of an electric vehicle is designed to withstand a certain maximum number of energy throughputs during its total life-span. Electric vehicles utilising high average power levels over long periods of time, e.g. trucks and buses, will reach this maximum energy throughput earlier than a passenger car mostly parked (for the same battery). Therefore, it is important to understand and evaluate operational conditions of the complete vehicle before the degree of electrification and subsequently the selection of battery is made.

In the development process of electric vehicles, and during the selection of degree of electrification, simplified or standardised duty cycles are often used. The same cycles are also used when developing the corresponding batteries. Based on duty cycles obtained from real driving, simplifications have been made in order to obtain artificial cycles having formats focusing on power trends rather than detailed and specific charge and discharge conditions. Even if no specific electric vehicle will operate exactly as these cycles, the cycles represent an average vehicle. An illustrative example of a tentative and simplified duty cycle is shown in Figure 5.3. This cycle constitutes a repetitive section and a deviant section, and could, for example, represent city driving followed by highway driving.

One of the advantages of these cycles is the possibility to compare different types of vehicles and battery designs. Other advantages are the easy implementation of these cycles in the control system of battery laboratory test equipment and that they facilitate correlations to be made between simulations and real test results. The drawback is that no vehicle will ever operate according to these cycles. Therefore, several different cycles are required to be used in the design process to increase the robustness of the system.

5.1.3   Energy and power requirements

How to select the battery for different kinds of vehicles is a process, or even an art, dependent on several factors, e.g. type of vehicle, performance requirements, packaging

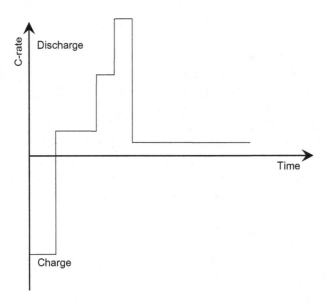

Figure 5.3 Schematic illustration of a tentative and simplified duty cycle.

constraints, and market. As for many other battery applications, there is rarely one size and solution suitable for all applications. From a vehicle perspective, the battery should obviously deliver and accept power according to the usage conditions, but the battery should also be robust and reliable, exhibit durability, and require minimal service and maintenance. Even if the usage of the battery in different types of vehicles varies, some general battery characteristics can be set, which will be discussed in the forthcoming sections.

The power and energy demands of various types of vehicles are roughly illustrated in Figure 5.4. There are large differences between the vehicle types and they therefore request different battery performance characteristics.

The *power-to-energy ratio* (P/E ratio) is a general design parameter. An HEV requires high-power, but has modest energy requirements, while an EV requires high energy, and the requirements of a PHEV are found somewhere in between. The ratio is influenced by vehicle weight, acceleration performance, vehicle top speed, and all-electric range. Higher ratios indicate shorter all-electric ranges, and vice versa. A difference of one order of magnitude between an HEV and a PHEV is not impossible, as illustrated in Figure 5.5. Normally, the average power and usable energy is used. Since a cell can be designed for either high-power or high energy (Section 4.3), the P/E ratio can be used to select the best cell for a given application.

The energy itself is often specified as the total amount of available or usable energy, which is not the same as the total amount of installed energy. The available energy is the energy corresponding to the SOC range where all the discharge and charge power targets are met. Depending on the type of vehicle, a limited SOC range is often used in order to secure power capability and prolong battery life. There is a maximum power level the battery should be able to deliver and accept, and that level should be

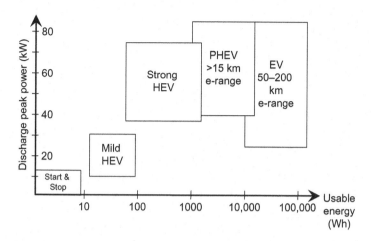

Figure 5.4 Power and energy demands for different electric passenger cars.

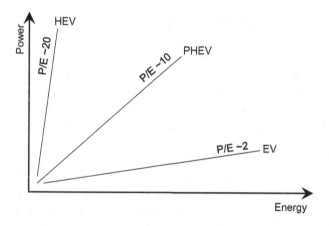

Figure 5.5 Power to energy ratio for EV, HEV, and PHEV, respectively.

independent of the SOC level. The power capability, of both charge and discharge, does however, depend on the SOC level of the battery: at high SOC levels, the deliverable power is high; at low SOC levels, the ability to accept power is high, and vice versa. An example of the power and energy requirements is illustrated in Figure 5.6.

There is a strong correlation between the available energy and the power performance requirements. Higher power levels than required can be achieved if the SOC range, and consequently the available energy, is diminished. Moreover, depending on the average SOC and SOC range, the battery can be designed to have different power capabilities for charge and discharge. The power capability of a battery is thus not a constant value, and will furthermore vary depending on temperature, C-rates, and the age of the cells. Many battery technologies have a narrow temperature range, often between 20–40 °C, for optimal performance. At lower temperatures, the power capability, especially for charging, is reduced or, at worst, the battery cannot be charged at all due to safety

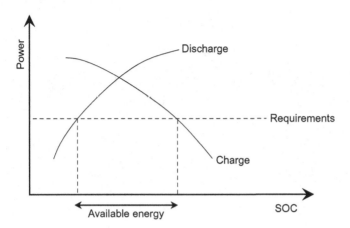

Figure 5.6 Power vs. SOC; the available energy in the range within charge and discharge targets are met.

issues. Durability is mainly affected by elevated temperature operation. The temperature dependency will be further discussed in Section 6.1.2.

The cells are optimised for either energy or power, and maximum C-rates for charge and discharge are determined by the cell chemistry and electrode design. How the battery has been used, the usage history, will affect to a considerable extent future performance. Therefore, limiting and monitoring the performance reduction during battery life is one of the most important battery design and control parameters and will be further discussed in Chapters 6 and 7.

The SOC range to operate in depends on the type of vehicle and the operation conditions. In general terms, a longer all-electric driving range requires a wider SOC range. There are mainly two operational modes with regard to the SOC level and range: *charge-sustaining* and *charge-depleting*. In the charge-sustaining mode, the SOC level is kept within narrow limits and the battery has to be charged or discharged in order to stay within the specified SOC range. On the other hand, in the charge-depleting mode, and as long as the SOC is high enough to deliver acceleration power, the SOC will gradually decrease during operation. The charge-sustaining mode is mainly used for HEV and PHEV applications, and the charge-depleting mode for the all-electric driving in EV and PHEV applications. Figure 5.7 shows a charge-sustaining sequence of operation and Figure 5.8 shows an operational sequence for a PHEV, including the charge-depleting mode.

In an HEV, the charge-sustaining mode is controlled within a narrow SOC range in the middle of the total SOC range. Sometimes, the SOC range is extended towards higher or lower SOC levels due to driving demands, e.g. hilly routes or extreme ambient temperature conditions. For the PHEV and the EV, the charge-depleting mode ranges from high to low SOC levels. The reason not to use the full SOC range is to maintain battery life and to always keep some energy left in the battery in case of an emergency. Like the HEV, the PHEV operates in a charge-sustaining mode as well. To fully utilise the key features of the PHEV, the charge-depletion mode is used as long as possible, i.e. as long as the SOC is high enough to deliver the required discharge power.

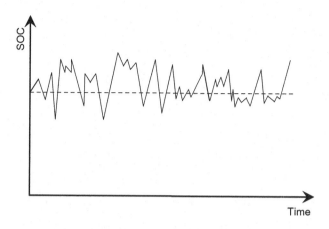

Figure 5.7 Charge-sustaining operational sequence.

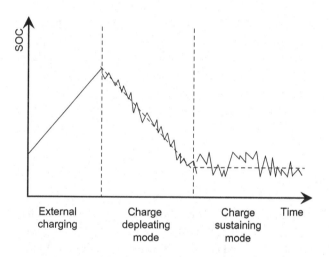

Figure 5.8 Operational sequence for a PHEV.

When charged externally, the SOC level in the battery is high and the vehicle can be operated in the charge-depletion mode until a specified lower SOC level is reached. Thereafter, a charge-sustaining mode is used and is essentially the same mode as the operational mode for an HEV, but typically at a lower SOC level. These SOC levels can vary from manufacturer to manufacturer and also depend on the cell chemistry.

How to manage the charge-sustaining and charge-depleting modes for optimal energy usage in the HEV and PHEV, respectively, varies from manufacturer to manufacturer and depends on the type of battery used. An EV is designed to manage the charge-depleting mode in an effective manner in order to keep the all-electric range as long as possible.

As can be understood from the above, the SOC range and level are key design parameters for all types of electric vehicles. Within the given SOC range, the power

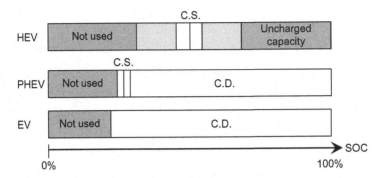

Figure 5.9 SOC ranges for different types of electric vehicles: HEV, PHEV, and EV.

requirements will be fulfilled for the range of the available energy (Figure 5.6), and the characteristics will vary among the different types of vehicles. In Figure 5.9, the utilised SOC ranges of the HEV, PHEV, and EV are summarised.

The SOC ranges of the EV and PHEV have much in common, while the ranges of the HEV stand out. The HEV battery is never fully charged and the charge procedures are interrupted at a pre-set lower SOC level. The SOC range set for an HEV is often in the order of 15–30% of a mid-point value. In the same way, an HEV battery is never fully discharged, and therefore all available energy is not utilised. This is also true for the batteries of EV and PHEV applications, but the relative amount of unutilised energy is smaller as more of the SOC range is used.

5.2   Battery design

Once the degree of electrification has been chosen, based on type of vehicle and usage conditions, the next step is to select the battery technology that can fulfil the performance requirements, and design and construct a battery to be installed in the vehicle.

Since there are a great many parameters to be considered when designing a battery for a specific vehicle, it is impossible to predict which battery technology is the most optimal. Therefore, in this chapter the battery design process will be discussed in general terms, i.e. the knowledge about important design steps will be described rather than a design manual provided.

A general route for selecting battery technology based on vehicle and environmental requirements is illustrated in Figure 5.10. The vehicle requirements will put constraints both on the power and energy performance of the battery. As described in the previous sections, usage conditions, such as the all-electric driving range and operational temperature ranges, are inputs for battery performance. During vehicle operation, losses in terms of drag and roll resistances will affect the power requirements of the vehicle. Moreover, energy management strategies in HEVs and PHEVs, i.e. the interactions between the electrical motor and the ICE, will determine the power demands of the battery.

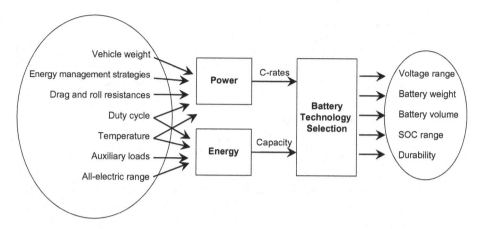

Figure 5.10 General work procedure to select battery technology and battery design constraints.

Power demands will result in the maximum and average C-rate the battery should fulfil, both for charge and discharge conditions. Required capacity, both installed and usable, will be the result of the energy demands. Power and energy are, however, not the only important parameters for an electric vehicle; how fast the energy can be utilised, i.e. the C-rates, is equally important, and Ragone plots (Section 1.6.4) are useful tools when matching battery characteristics with vehicle requirements, and will be discussed further later. Moreover, temperature specifications will have a considerable effect on power and energy demand, and thereby the optimal battery technology for the intended vehicle. The cell selection will strongly affect performance and durability, as well as weight and size of the battery.

One of the fundamental aspects of battery design is consistent performance, i.e. it may vary from a long-term perspective or due to varying ambient conditions, but should be consistent from the short-term perspective. Not only performance will be influenced by the cell selection and battery design constraints, the reliability and safety are also affected to a large extent. The battery design should also include protective devices and features that may prevent, or at least minimise, possible damage, such as temperature sensors and thermal control, charge control, and overcharge protection fuses.

The understanding of the possibilities and limitations of the cells, as well as the knowledge about how energy and power demands affect the battery design constraints, provides constraints and guidelines for battery development. In order to produce a battery ready for usage in an electric vehicle, more is required than just connecting cells in series and/or parallel. The next step will be to translate the knowledge about battery performance into algorithms to be incorporated in the battery management system, and is a task combining different skills: e.g. control, thermodynamics, and electronics. These functions, e.g. power and energy management, temperature range, and thermal management, voltage and current regulations during charge and discharge, will be discussed in Chapter 6. In Chapter 7, battery construction from cell selection to battery design will be discussed, and the different battery technologies suitable for electric vehicles will be briefly compared.

5.2.1   General design criteria

There are, however, some general criteria to be considered even before setting the cell and battery design constraints. These criteria, listed and described below, will in most cases influence the properties of the subsequently selected cells.

5.2.1.1   Mechanical stability

Mechanical stability can be evaluated and interpreted at different levels. Mechanical stability of a battery is related to the integration of several hundreds of cells into a battery requiring high mechanical stability, from cell-level to the complete battery construction. Cells are connected in series and/or parallel and welding is a commonly used method to create connection stability. The cell-to-cell connections should also enable electric connections between the cells and limit current leakage.

In a vehicle application, vibration is one of the most critical aspects of the battery construction when it comes to mechanical stability. Therefore, it is important to use a cell design unaffected in its performance by vibrations and to have cell-to-cell connections able to withstand the vibration forces. Also the positioning of the battery in the vehicle affects the vibration levels.

Corrosion of different mechanical parts of a battery is another design parameter. The humidity and the presence of, for example, salt and dust in the air will cause a corrosive atmosphere affecting, for example, cell-to-cell connections, contactors, and electronic circuit boards. Moreover, a car wash may severely affect the battery depending on the integration and position in the vehicle.

Since vehicles are also occasionally subjected to impacts and crashes, the mechanical integration and encapsulation of the battery must be properly considered, as well as the position of the battery inside the vehicle. Here, statistics are used to detect vehicle zones more vulnerable, and thus less suitable. All issues taken into account, the mechanical stability of a battery is very much related to its position in the vehicle.

5.2.1.2   Temperature

Since the cells are affected to a large extent by temperature, the temperature in which the electric vehicles will run sets some boundary conditions for battery design in order to secure performance, safety, and life. For example, the climate of different regions around the world will influence how the thermal management system of the battery is designed. In some areas, heating of the battery may be required during parts of the year and in other areas the battery needs substantial cooling. In many areas, the temperature can be above 40 °C and in a parked car the temperature may substantially exceed that temperature. On the other hand, in cold climates the vehicle should still be able to start and operate at temperatures below ☐40 °C. The temperature specifications of a battery for electric vehicles may range from ☐50 °C up to +80 °C and are challenging. Often the specifications are divided into operational and storage temperatures. At high and low temperatures, there are cell processes affecting the performance in a negative way and usage outside the 'safe' range can damage the cells and thus the battery severely. In addition, temperature is not only a parameter affecting performance, but is also a safety parameter to consider. A battery, subsequently, always includes some kind of thermal management to keep the battery temperature in an optimal interval for the cell selected.

5.2.1.3   Battery life

In most cases, it is not possible or profitable to exchange the battery during the life of the vehicle and therefore the battery must be designed for a long life. The *end-of-life* (EOL) conditions are reached when the battery fails to fulfil the application requirements; in an EV the capacity fade is the determinant factor, and in an HEV the power fade. Often the limit is set when the capacity of the battery is 80% of the *beginning-of-life* (BOL) capacity. This 80% limit is commonly accepted within the battery community as the EOL criteria, and batteries for electric vehicles are normally designed with that in mind. Obviously, the battery is still usable below this limit and from a driver's perspective a decrease of 20% in capacity is often unnoticeable since most batteries can still deliver the requested power.

All cells undergo performance degradation due to irreversible side reactions taking place during the discharge/charge cycles and under storage. At cell level, this will be observed as capacity and power fades. There are two main types of cell degradations: *calendar* and *cycle life*. The calendar life is defined as the number of years a specific battery will last and cycle life is defined as the number of charge/discharge cycles the battery could perform for a specified set of operating conditions. The calendar life refers to batteries incorporated in the intended application and is affected by the degradation mechanisms occurring in a cell whether used or not. Passenger cars, parked most of the time, are more dependent on the calendar life, and vehicles frequently used are dependent on the cycle life.

All batteries have a limited *shelf-life*, i.e. the time the battery can be stored without severe degradation without being incorporated in the intended application, and this has to be taken into account, for example, at the vehicle manufacturing plants or in spare part storage. Shelf-life is related to *self-discharge* (Section 7.1), but differs slightly as the battery is not yet being implemented in an application. Battery degradation aspects will be discussed in detail in Chapter 7.

5.2.1.4   Voltage and cycling characteristics

The cell performance must be matched with vehicle performance requirements in order to fulfil the optimal operation of the vehicle. The ability to deliver and to accept power according to specifications is important in all driving situations. In many applications, a constant voltage level of the battery, i.e. a flat discharge profile of the cell (Section 1.4.2), is preferred in order to maintain the same performance over a wide SOC range.

In an electric vehicle, however, the SOC level is an important control parameter for predetermining the energy management of the vehicle and to calculate the remaining driving range accurately. In such cases, a sloping discharge profile or a flat profile, which contains well-defined voltage steps, are more interesting alternatives. How the SOC is determined is further discussed in Section 6.2.1.

Usage of, and access to, EVs and PHEVs is affected to a considerable extent by the external charging time, e.g. fast charging. The cell reactions can, however, not be stressed without long-term damage to the cells since the kinetics and reaction routes are often time dependent. Fast-charging procedures (i.e. charging at high C-rates) may therefore result in shortened battery life. The side reactions occurring in the cells due to, for example, overpotentials caused by non-optimal charging conditions, and the corresponding losses, must be limited or avoided. Charging procedures are described in Section 6.1.1.

5.2.2   Cell selection

In the next step of the battery design process, the most suitable cell is selected and combined in series and parallel configurations in order to fulfil the performance requirements. Usually, cell selection is based on extensive testing where the performance of the cell has been evaluated according to the intended usage. This is crucial as the performance of a cell, when assembled in a particular battery environment, can be significantly different from that of an individual cell. The cells themselves affect each other, as well as the true operating temperature. Specifications and data sheets provided by the cell manufacturer should be used merely as a rough guide as it is not always possible to extrapolate the data properly to determine the performance of a battery having cells in series and/or parallel configurations and operating according to specific usage conditions. Moreover, the operational constraints, and their variations, like the charge/discharge profile, and ambient temperature influence the final performance characteristics of the cells.

The cell format is a factor determining size, volume, and shape of the final battery design. There are mainly three cell formats of interest for electric vehicle applications: cylindrical, prismatic, and pouch (Section 4.4). The cylindrical cell, wound centrosymmetric, is the most common cell format for power-optimised cells with the advantages of easy manufacturing, high mechanical stability, and usually good cycling stability. From a packaging perspective, cylindrical cells make insufficient use of space with low packaging density, but the cavities can be used for cooling purposes. In a prismatic cell, the electrodes are either stacked, wound, or folded, providing better space utilisation. A pouch cell is of soft-case prismatic geometry and makes the most efficient use of space by eliminating the stiff casing. The soft casing reduces weight, but lacks the mechanical protection of cylindrical and prismatic cells, which is why these cells often require some external mechanical support.

The size and format of the cells depend on manufacturing processes, targeted capacity, and area of use. For example, the battery will be larger and heavier if a cell optimised for energy is used in a power-demanding application, as will be illustrated by some examples later in this section. As can be understood, cells from different sources (e.g. manufacturers and batches) should preferably not be mixed in a battery without careful control and matching of the cells before battery assembly is needed to avoid balancing constraints. Differences within the battery are often amplified by degradation, and small variations at the beginning of life can grow large and, as a consequence, the end of life can be reached faster than expected. Cell imbalance can also occur after battery assembly, mainly due to thermal gradients within the battery, and corrective actions must be taken to prevent an accumulation of imbalance. It is, however, important to use the same cell technology and cell type throughout the battery to secure balance between the cells and to limit abuse situations.

All cells within a battery are not equal throughout even if they are based on the same battery chemistry and have the same cell format, manufacturer, and nominal capacity. Cell-to-cell variations, which can, to some extent, be characterised at the cell level, will affect battery performance considerably. The properties of the active and non-active materials of the cell play a significant role in the overall cell performance and result in differences often observed as small variations in cell capacity. The amount of active material, tap density, and resistance offer some ways to examine the cell-to-cell variations, even for cells having the same rated capacity and originating from the same production batch. The intrinsic properties of the materials can be equal in many aspects (particle size, morphology, microstructure, etc.) and still give rise to performance variations. For example, the electrode processing will affect the performance, and different electrode morphologies can have different rated capacity, in the long run resulting in wide deviations in cycle life between cells.

The typical battery designed for electric vehicles exhibits cells connected in series, as well as in parallel in order to fulfil the performance and durability requirements. Cells in series are needed to achieve the desired operating voltage, each cell adding to the total voltage. Capacity, though, does not change with the increasing number of cells since the same current passes through all of the cells. On the other hand, cells connected in parallel will have the same voltage as a single cell, but capacity will increase with the number of individual cells since the current, in this case, is the sum of all currents passing through each individual cell.

Thus, series-connected cells provide high voltage but low current, and parallel connected cells provide high current but low voltage. Combined series and parallel configurations allow design flexibility to meet the vehicle requirements, i.e. the ability to deliver energy and power according to the operational conditions. Individual cells are, however, limited by the capacity and C-rates. The number of cells connected in series is therefore based on the lower cut-off voltage of the cells. Regarding cells connected in parallel, the average discharge current integrated over time times the cell capacity is the measure used to determine the required number of cells.

Many batteries are made of cells divided into modules in order to have a robust battery in terms of, for example, manufacturing, packaging, control, thermal distribution, maintenance. Each module consists of cells connected in series and/or parallel. Module configurations are commonly described by their number of cells in series and parallel; 8S2P would refer to eight cells in series and two of these strings in parallel (16 cells in total). A module of 16S4P would, hence, have twice the voltage and capacity of the 8S2P module. The modules can then be connected in series and/or parallel to compose the complete battery.

5.2.2.1   Technology selection

Based on the cell selection discussed above, the required cell performance has to be fulfilled at the atomic level by the cell reactions for any specific battery technology. In order to select the most optimal cell for the given task, some general technology comparison has to be done. The most suitable technologies and concepts for electric vehicles are described in Chapter 2, and how Li-ion cells are optimised for different performance characteristics is described in Chapter 3.

To be able to select cell technology for a specific type of electric vehicle and usage conditions, it is important to understand the variety of materials, cells, and batteries available and their characteristics. This includes, for example, the energy and power relationship, how the cell impedance varies by temperature, the pulse discharge capability as a function of temperature and load, the charge/discharge characteristics, and how material degradation is affected by the temperature distribution caused by packaging constraints. Simply comparing the listed performance supplied by cell manufacturers is a starting point, but never enough since these data are often generalised, and to a large degree based on optimal cell designs and operating conditions. With this in mind, the cells must be tested according to the actual operational conditions and with a proper fundamental understanding of the underlying electrochemical and chemical mechanisms.

There are a number of design possibilities since the performance characteristics vary widely. Cells having the same capacity, but based on different technologies, can behave differently. For example, the capacity can be the same for a NiMH cell and a Li-ion cell, but the rate capability can be noticeably different between the two. Two cells of the same technology and capacity can also perform differently, e.g. two Li-ion cells of the same rated capacity, but with different active electrode materials will exhibit different cell voltage levels.

When comparing different technologies, it should be noted that cell design, chemical composition, and manufacturing conditions also all significantly affect performance. There are, however, general comparisons that can be made:

- cell potential and discharge profile
- energy storage capability, in terms of Wh/kg or Wh/l
- charge acceptance and discharge rates
- life: calendar and cycling, time and performance dependency

The technology selection often begins with looking at cell potential and the character of the discharge profile, as this will indicate the number of cells needed and what voltage range the electric drive system has to be able to handle. Figure 5.11 schematically shows how the voltage changes as a function of SOC for four technologies: Pb-acid, NiMH, Li-ion, and EDLC.

There are some major differences; most obvious are the large variations in voltage level, but also the shape of the profiles. The discharge profiles for the three battery technologies are flat over a wide SOC range, and at the end of the discharge the potential drops rapidly. On the other hand, the EDLC exhibits a potential decrease linear with SOC.

The next parameters in line to be considered are energy density and specific energy, reflecting weight and volume of the final battery. As for the discharge profiles, the energy density varies with the chemistry and the materials in the cells. Pb-acid cells

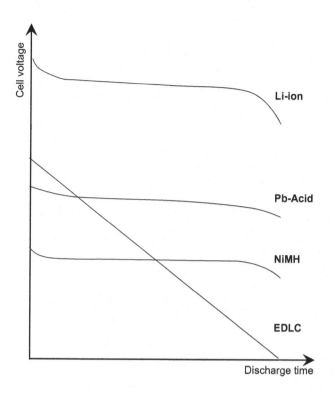

Figure 5.11 Discharge profiles for Pb-acid, NiMH, Li-ion, and EDLC technologies as a function of SOC.

will, for example, always be heavy due to the density of lead. Figure 5.12 illustrates the relationships between energy density (Wh/l) and specific energy (Wh/kg) for the same technologies. The trends are valid for batteries, as well as for cells.

The power density, on the other hand, is mainly affected by the cell design, e.g. particle size, morphology, and electrical contacts inside the cell, i.e. all that minimise cell resistance and impedance. Therefore, it is not possible to calculate a theoretical value of the battery for the various technologies.

Depending on the degree of electrification (i.e. HEV, PHEV, or EV) and the type of vehicle (e.g. passenger car or truck), the different battery technologies are more or less suitable. This is mainly due to their different rate capabilities, which can be the determining step in the technology selection process. Modified Peukert plots (Section 1.6.3) and Ragone plots (Section 1.6.4) are key tools for rate capability comparisons. However, even HEVs of the same type and degree of electrification can be equipped with different battery technologies due to differences in duty cycles and energy management strategies. Therefore, it is difficult to determine the optimal battery technology of a specific vehicle alone, but it can often be concluded that some technologies are less suitable.

In the Ragone plot of Figure 5.13, some generalised vehicle requirements for EVs, HEVs, and PHEVs are shown together with indicative power and energy densities of the specified technologies.

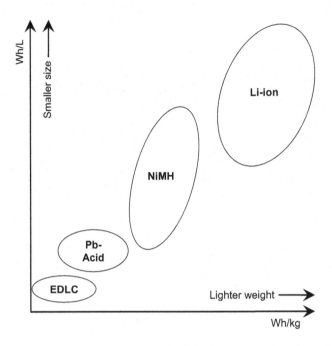

Figure 5.12 Energy density (Wh/L) versus specific energy (Wh/kg) for the Pb-acid, NiMH, Li-ion, and EDLC technologies.

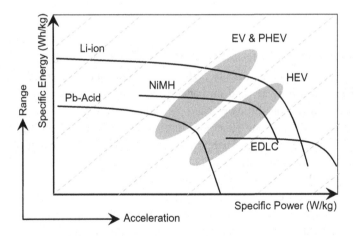

Figure 5.13 Ragone plot of different battery technologies suitable for vehicle usage and indicative passenger car requirements for EV, HEV, and PHEV.

The main difference between the different types of vehicles is the charge/discharge rates, i.e. the power capability the battery must be able to provide. The vehicle requirements illustrated in Figure 5.13 refer to passenger cars only, and, as expected, the power requirements are of the same order of magnitude. The divergences are found in the rates of power delivery and in the energy density, i.e. the all-electric driving range. As indicated in the Ragone plot, the power capability highly affects the energy performance of the battery; if high C-rates are required, the all-electric driving range will be drastically reduced. On the other hand, trucks and buses require different performance characteristics, and will differ to a large extent in power, energy, and rates.

5.2.2.2 Example: cell selection consequences

Selecting the right, or wrong, cell for the intended electric vehicle configuration may result in oversized batteries, being unnecessarily heavy and expensive. The energy and power-optimised Li-ion cells described in Section 4.3 will be used to illustrate the battery design both for an HEV and an EV, using optimal and non-optimal cells.

The boundary conditions are a passenger car with an electrical drive system requiring a nominal voltage of 300 V, and the ability to use 100 A as the maximum current. The assumed peak-power is 60 kW both for the HEV and the EV. The power-to-energy ratio (P/E ratio) for the HEV is 20 and for the EV 2, both according to the data in Figure 5.5.

The Li-ion cells at disposal are a 5 Ah power-optimised cell and a 25 Ah energy-optimised cell, each made of one of the three different active positive electrode materials: LMO, NMC, and LFP. The maximum C-rate for the power-optimised cell is assumed to be 20C, and for the energy-optimised cell 4C. In both cases, a maximum current of 100 A can be used (i.e. 5Ah 20C and 25Ah 4C), hence equal to the maximum current demand of the electric drive system. The maximum power that one string of series-connected cells (300V) can deliver/accept is 30 kW (i.e. 300 V × 100 A). The voltage does vary within the SOC range, but ideal conditions are used for illustration.

First, the required number of cells connected in series to reach 300 V depends on the cell voltage: LMO 4.0 V, NMC 3.7 V, and LFP 3.2 V, respectively. This results in about 75 cells of LMO, 80 cells of NMC, or 90 cells of LFP.

Secondly, the energy content of one string of cells is:

$$Energy = number\ of\ cells \times energy\ density\ per\ cell\ (Wh/kg)$$
$$\times cell\ weight\ (or\ voltage \times capacity)$$

Based on the data in Tables 4.1 and 4.2, Section 4.3.3, one string of series-connected cells will result in about 1.5 kWh for the 5 Ah cell and 7.5 kWh for the 25 Ah cell.

Now it is time to evaluate and summarise the number of cells in series and parallel configurations required to fulfil the vehicle performance requirements (summarised in Table 5.1).

In the HEV case, two parallel connected strings of series-connected cells will fulfil the power and energy requirements by using either the 5 Ah or the 25 Ah cell. In the case of the 25 Ah cell, the system will be oversized in terms of energy; it is impossible to get more than the exactly needed power out of the battery, while there is more energy than needed.

The EV power requirements, on the other hand, are also fulfilled by the two strings of series-connected cells. The different cells show large deviations; 10 strings (3 kWh per string) in parallel are required to fulfil the energy requirements in the case of the 5 Ah cell, while four strings are satisfactory for the 25 Ah cell.

Table 5.1 Power and energy optimisation based on cell sizes.

	HEV	EV
P/E ratio	20	2
P req. (kW)	60	60
E req. (kWh)	3	30
Number of parallel strings to fulfil P:		
– based on 5 Ah cells	2	2
– based on 25 Ah cells	2	2
Number of parallel strings to fulfil E:		
– based on 5 Ah cells	2	20
– based on 25 Ah cells	1	4
Summary: To fulfil both P and E	2	4

Table 5.2 Chemistry and cell size optimisation of weight and cost.

	Cell	Chemistry	Number of cells	Total cell weight (kg)	Total cell cost (CU)
HEV	5Ah	LMO	150	26	171
		NMC	160	32	**149**
		LFP	180	25	205
	25Ah	LMO	150	95	239
		NMC	160	86	274
		LFP	180	104	248
EV	5Ah	LMO	750	131	855
		NMC	800	160	744
		LFP	900	126	1026
	25Ah	LMO	300	189	**477**
		NMC	320	173	547
		LFP	360	209	497

The results of this rough example above clearly indicate the importance of selecting the right cell. This will be further illustrated in the next example of battery weight and cost, in Table 5.2, also this with cell chemistry variation. The cell weights are taken from Tables 4.1 and 4.2, and the cell cost is normalised to the energy density (100 Wh/kg here corresponds to 1 cost-unit (CU)). The most cost-effective solutions are highlighted. In summary, from a cost perspective, these examples indicate that the small cells (5 Ah) are suitable for HEVs, while the larger cells (25 Ah) are suitable for EVs.

The examples above are just rough estimations, and should be seen as an illustration of the complexity in selecting the appropriate cell and chemistry for different vehicle applications. The P/E ratios used are only indicative numbers, and will differ from more precise calculations when there is a specific vehicle in mind. Moreover, the examples are based on one specific cell geometry where merely the active material of the positive electrode has been varied; all other parameters have been fixed. Changing any of the cell parameters, e.g. cell format and housing material, amendments in the electrolyte, and/or the negative electrode, will change the weight and the energy density of the cell, and the results presented in Tables 5.1 and 5.2 will change accordingly. The only fixed value is the cell voltage, which will remain constant as long as the same negative electrode material is used. To complicate the picture even further, an energy-optimised cell in an application requiring even higher P/E ratios, e.g. a city bus, could be added, giving a quite different picture.

It should also be emphasised that the data of weight and cost presented in Table 5.2 refer only to the cells. The battery weight and cost will be substantially different since the battery consists of more components as described below, and particularly the control level chosen for the battery management system often carries a substantial part of the total battery cost.

5.2.3   Additional battery components

A battery for electric vehicles is more than a number of cells connected in series and parallel and embedded in a protective mechanical housing. In addition to the cells, several other components are needed to complete a battery: components to secure performance, durability, and safety, during vehicle operation, as well as when parked or in service. The performance control of the battery is mainly secured by the battery management system, including battery control and management, which will be discussed in Chapter 6. In the following sections, the thermal system, the electronics required, and how the actual battery design and construction can affect the overall battery safety will be described. It should be noted that each component added increases the weight, size, and complexity of the battery, and consequently the complexity of the battery installation and the final vehicle will increase.

5.2.3.1   Thermal system

One important design issue is the temperature distribution within the battery. The main purpose of the thermal system is to keep the cells within the preferred temperature range to secure optimal performance at all ambient and operational conditions, and to ensure a low cell-to-cell temperature difference. Cooling and heating of the battery, or parts of the battery, may be required depending mainly on the ambient temperature. The temperature requirements are furthermore often divided into operational and storage temperatures.

Depending on vehicle type (e.g. HEV or EV) and battery position, the temperature in the engine compartment, the passenger compartment, or the cargo space can vary significantly, resulting in different thermal solutions. Moreover, splitting the battery into modules and placing them in different positions within the vehicle in order to spread the mechanical load or to fit more battery capacity in, may result in various temperature conditions. On the other hand, if the battery is kept as a single large unit, there will most likely be a temperature gradient through the battery. The cells closer to the edges will sense the ambient temperature, which may be cooler, while the cells surrounded by other cells will be warmer unless coolants are provided to remove the

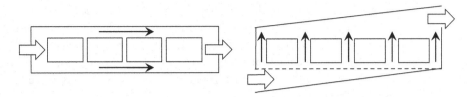

Figure 5.14 Possible air-cooling configurations.

heat. The temperature is measured on the skin of the cells, preferably in between the cells, or in the cooling channels, i.e. measuring the temperature of the cooling media at different positions. Besides cell selection, temperature equalisation across the battery is the most important design tool to minimise cell variations and secure performance requirements, and in the end to prolong the life of the battery.

In general, there is no perfect cooling or heating concept for all battery technologies or all types of electric vehicles. Different cooling and heating strategies are needed due to the variations in usage conditions, physical size of the battery, packaging constraints, comfort and safety aspects, and ambient environment. Cell chemistry and format will also affect the thermal system. For example, a prismatic cell can obtain cooling or heating at different sides of the cell, in contrast to a cylindrical cell, and the thermal capacity of the mass of the battery may affect the cooling and heating requirements. Heat dissipation is therefore an important engineering task for the best possible battery design.

Furthermore, there are different cooling and heating media to consider, e.g. air, liquid, refrigerant, and their capabilities are diverse; while a refrigerant medium offers proper cooling, but limited heating capability, air and liquid media are useful both for cooling and heating purposes. Air-cooling may, however, be limited due to hot ambient conditions, and especially if the temperature difference is low.

The air-cooling or heating can be facilitated by either passive or active means, where active air-cooling is sometimes called forced air-cooling. Depending on the placement of inlet and outlet, the forced air will pass the battery in pre-defined pathways. Figure 5.14 illustrates some possible set-ups for air-cooling.

The configurations of the cooling pathways within the battery are designed depending on the inlet air temperature and the heat generated due to battery usage. In the left configuration of Figure 5.14, the main air-flow passes the top and bottom of the cells, where the electronics often are placed, while only a minor portion of the air will pass in between the cells. In the right configuration of Figure 5.14, the air will flow in between the cells, but this also requires more space.

Liquid cooling is widely used, especially for batteries utilising energy-optimised cells and batteries in high-power-demanding applications, i.e. in applications where the heat generation from the cells is extensive and more problematic.

The temperature specification of a battery for electric vehicles poses a challenge as the battery should be operational even when the ambient temperature varies from $\Box$50 °C up to +80 °C. This will of course greatly affect the thermal management of any vehicle.

An EV used in cold climates, i.e. ambient temperatures at about 10 °C or lower, often has an external heater installed to keep the temperature at acceptable levels both in the battery and in the driver compartment. In HEV and PHEV applications, it might be possible to design the thermal system of the battery in combination with other thermal systems of the vehicle, e.g. the cooling circuit of the ICE, or the electric drive system might be utilised for thermal control of the battery, all depending on vehicle integration. This may facilitate vehicle installation and reduce the number of components installed, but must be seen in perspective of the complete vehicle in order to secure full functionality.

Even though the thermal system also acts as a safety protection device, in some cases it can actually cause severe situations. Depending on the application, the heat generated by battery usage may be employed in the vehicle climate control system to heat the passenger compartment. In case of cell venting, though, the emitted gases from the cell must not come into contact with the passengers. If a liquid cooling medium is used, it is important to prevent any contact between the electrical system and the cooling medium since any leakage may be hazardous.

5.2.3.2   Electronics

All battery technologies require various degrees of protection, often by electronic protection devices to keep voltage, current, and temperature within their respective preferred limits. For example, as Li-ion cells are sensitive to overcharge and over-discharge conditions, voltage and current sensors are usually applied for monitoring and protecting the cells, and consequently the battery. This could include devices to detect if the current limits have been reached in order to interrupt the circuit, like CID (Section 4.4.4). Current is, however, difficult to measure, so this is performed by measuring the voltage across a low-ohmic resistor, series-connected in the current path. If the current limit has been reached, the sensing circuit interrupts the current path by triggering a switch, e.g. a semiconductor device or a relay.

Fuses are often used for protection and there are several options to consider. Using a thermal type fuse, the battery will be shut down if the temperature exceeds a predetermined limit and the battery can be used again only after the fuse has been exchanged. Moreover, re-settable fuses with similar functions as the thermal fuses are often used. After opening, the fuse will automatically reset as soon as the fault conditions have been removed, e.g. the battery has reached a temperature within the operational boundaries. A temperature rise can be caused by resistive self-heating due to high currents passing through the fuse or by convection of the ambient environment; the fuses detect both high currents and non-optimal temperatures.

There are other circuit devices, such as thermistors having a resistance that varies with temperature. As for thermistors with a positive temperature coefficient, the resistance will increase gradually with temperature and over a limited range a linear relationship exists. The opposite is true of thermistors with a negative temperature coefficient; the resistance decreases with increased temperature. The thermistors can be used to control the thermal management of the battery or to terminate charging procedures if an unexpected temperature rise occurs. All electronic protection devices draw current from the battery, consequently reducing the efficiency of the battery. While not all possible electronic devices used for safety supervision in a battery are covered, the electronics related to cell balancing represent an essential part of the electronics in a battery, and will be described in some detail in the context of SOC estimations in Section 6.2.1.

5.2.4   Design impact on reliability and safety

The reliability is determined by the cell and battery design, as well as the manufacturing processes. Therefore, cell and technology selection, battery design and construction, together with well-defined battery monitoring and protection devices, are of key importance for optimal, reliable, and safe operation of the battery. Reliability and safety are also determined by the interactions between the cells, the electronics, and the mechanical structure of the battery, as well as the management system. Cell selection and purity of materials, and cell capacity all have an impact on battery reliability and safety. Small cell variations at BOL will, during battery life, become pronounced, and may grow large enough to cause abuse situations if not handled properly. Impurities inside the cell can lead to internal short circuit and/or accelerate side reactions, e.g. pressure build-up, which, at worst, may inflict thermal runaway (Section 6.1.2.1). Quality control in the cell production plants is thus one of the first steps towards a reliable and safe battery for electric vehicles.

Most cells are sealed and gas pressure might build up inside due to side reactions, particularly during overcharge conditions. Therefore, the cells are often equipped with some kind of venting device in order to release the pressure in a controlled way, and to avoid the cell rupturing. Unfortunately, it is difficult to measure or monitor the internal pressure of the cells. During normal operations, however, cells are never meant to vent.

As manufactured, the cells may be equipped with other safety devices, which can disconnect the cell if failure is likely to occur. These devices can be incorporated into the cell design as external resistances, switches, or fuses. They mostly act as safety vents for internal cell failures and are often non-resettable. Some cell-protecting devices for Li-ion cells are described in Section 4.4.4.

The use of high-quality cells does itself not guarantee a safe battery, as there are mechanical issues which can cause failure mechanisms due to improper battery construction, e.g. poor cell-to-cell connections, lack of insulation, and insufficient case assembly. Regarding the mechanical integration of cells into a battery, there are a few general procedures to consider: the cell-to-cell integration, the encapsulation of the cells, the configuration and material for the battery housing, and the selection of the terminals and contactors. As described earlier, welding is a commonly used technique for cell-to-cell connection. The resistance of the welds should be as low as possible to minimise voltage losses and to increase overall battery efficiency. During the welding process, the heat generated can affect the chemistry of the cells to cause fast degradation or abuse situations at the parts of the cell closest to the welding points. The welding process, subsequently, is a critical step in designing a reliable and safe battery.

Except for internal cell and battery protection devices, the battery is usually equipped with protection devices to respond to external events. One example is high temperatures in other parts of the vehicle requiring power shutdown, e.g. in the case of a traffic accident or an electrical malfunction in the traction power system of the vehicle. In such cases, the battery needs to be isolated to evade battery damage and prevent the battery from inflaming the situation. Therefore, the battery is often equipped with a switch to cut-off the main current path, a switch triggered by external signals often called the battery disconnect unit. If a battery is short circuited through the external terminals, the chemical energy is converted to heat and hazardous situations may arise. In order to avoid that scenario, the terminals should be physically isolated. The battery often needs other protective means in case of external events, especially related to mechanical impact and crashes. Therefore, the position of the battery in a vehicle is an important design issue directly related to safety.

6 Battery control and management

The battery is merely an energy storage and the key for all-electric vehicles is understanding how to *use* the battery in the most optimal way in order to secure vehicle performance over a long period of time. The operating and controlling strategies of a battery rely on the understanding of the fundamental cell constraints, which are turned into battery and vehicle control strategies, and implemented as algorithms in the *battery management system* (BMS): the control unit of the battery. The BMS will control and monitor the performance and status of the battery and communicate the operational constraints currently available to the control system of the vehicle. There are many cross-dependent parameters to be understood and to be incorporated in a robust and reliable control system. Input data for the BMS are the state functions, e.g. state of charge and state of health, battery temperature, and usage history, required to secure optimal performance in a durable and safe manner. How this control and communication is handled depends on the battery and vehicle manufacturers, and is not covered in this book. Instead, the underlying fundamentals will be discussed in terms of electrochemical and material constraints. In the following sections, battery control and management will be described: charge control and methods, thermal and safety management, as well as the state functions, i.e. *state of charge* (SOC), *state of health* (SOH), and *state of function* (SOF).

6.1 Battery management system

The *battery management system* (BMS) utilises a number of parameters that are linked to each other and most of the key parameters are path dependent, and the usage and environmental history affects future operational possibilities. Each of these parameters affects the battery control and management system: temperature, voltage range, current, and energy throughput. *Temperature* is one of the most important parameters for the BMS and the corresponding control strategies. The battery should be used within a specific temperature range, a range defined by the chemistry inside the cell. At temperatures outside this predefined range, higher as well as lower, side reactions may take place, side reactions limiting battery life and possibly causing abuse situations. In a similar way, the *voltage range* is also determined by chemistry and operation outside this range may cause severe degradation and abuse situations. Another parameter having a considerable effect on battery performance, life, and safety is the *current* used, a parameter for which it is more difficult to set the boundary conditions, and a more complex relationship is found. Current is very important for an electric vehicle in order to fulfil the power requirements of all driving situations, and will vary more or less from second to second. The power status at any given moment, i.e. the current that can be accepted or delivered by the battery, is determined by the voltage of the battery; close to the upper cut-off voltage a low charging current but a high discharging current can be used, and vice versa at the lower cut-off voltage. These parameters will reflect the charge procedures, particularly for all-electric driving possibilities in EV and PHEV applications. *Energy throughput* will also reflect the battery functions and life. This parameter can roughly be translated into the odometer in a conventional car.

The dynamic behaviour of the vehicle requires a BMS much more advanced than required for a laptop or a cellular phone, which are only charged at periodic intervals and have low-current discharging usage conditions. An electric vehicle often requires fast and high current charging and discharging, especially in HEV applications, associated with dynamic power requirements. This rapid cycling condition of the battery requires sophisticated control strategies to regulate and control the performance requirements in real time, and at the same time secure performance and safety in combination with battery life protection. Therefore, an advanced BMS is required. The BMS is at the interface towards other control units of the vehicle and must act in real time according to the rapidly changing driving conditions.

The BMS combines the material and electrochemical constraints via parameterisation with dynamic driving conditions in a multi-disciplinary approach to control the charge and discharge currents in real time. Maintaining the battery in an accurate and reliable operational condition is one of the tasks for the BMS, and efficient battery management is needed to prolong the life and secure the requested performance of the battery in a safe manner. Moreover, another function of the BMS is to keep the battery within defined safety limits, which is often realised differently, depending on type of electric vehicle and manufacturer. The main functions of the BMS are to protect the battery, secure performance in a safe operational mode, and to act as an interface between the battery and the vehicle. Therefore, it will estimate and control the *state of charge* (SOC, Section 6.2.1), the *state of health* (SOH, Section 6.2.2), and the *state of function* (SOF, Section 6.2.3). Furthermore, the BMS is a unit for diagnosis and fault tracing, e.g. parameter detection, security tracing and alarm, charge control, and battery balancing. By controlling the dynamic character of the charge and discharge currents, the BMS limits the risk of overcharging and over-discharging, and keeps the cells and the battery within the pre-defined voltage range.

In HEV applications, the battery is continuously charged and discharged to fulfil the requested power demands. At every instant, the BMS must be able to decide whether the power, wholly or partly, can be accepted, or delivered, in order to keep the battery within the defined voltage limits, and communicate to the vehicle control unit in order to secure vehicle performance. The control strategies used may be determinant for optimal energy usage and the battery life. GPS data and other traffic information data can be used in the overall energy management strategies of the vehicle to optimise usage of the battery, especially in HEV and PHEV applications where the use of both energy sources should be optimised.

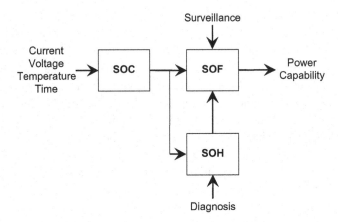

Figure 6.1 An example of work procedure of the BMS control unit.

In an EV, the BMS must also predict the remaining electric driving range in a reliable way. This is commonly achieved by model-based control strategies using the instantaneous information about currents, voltages, and temperatures as input data. Figure 6.1 shows an example of the work procedure of a BMS controlling the performance outputs of the battery. This model-based method requires preparatory work developing robust and reliable battery models based on extensive cell and battery testing.

The BMS will also control the long-term behaviour of the battery based on instantaneous and historical usage. This is done by employing the balancing procedures (Section 6.2.1.4) and controlling the temperature, voltage, and current levels. The overall goal is to prolong life and to monitor and raise an alarm about safety issues.

As there are many functions and requirements to be fulfilled, a number of devices are used to track and measure different parameters: e.g. sensors, actuators, protection electronic circuits, safety interlocks. Moreover, the BMS should also include diagnosis and service tools to identify any internal or external causes of battery failure and warn of any abuse situations. From a security point of view, the detection of malfunctions in the sensors used for voltage, current, and temperature, but also insulation failures, is essential. Abuse situation alarms, e.g. gas detection, are included in the diagnosis tools.

The BMS must protect the battery by fault detection in many situations, and complexity is to be considered carefully. In many applications, the battery is disconnected if abuse situations occur or are likely to occur. Disconnecting the battery electrically from the vehicle will protect the battery, the vehicle, and the surroundings, and will prevent the battery from unsafe currents. Many faults may result in the same observation, even when caused by different mechanisms: chemical degradations, physical cell and battery manufacturing defects, malfunctions in the electronic protection devices and in the control units.

To perform fault detection, usage history is valuable, sometimes crucial, and it preferably includes current and voltage history, as well as environmental conditions, such as temperatures, vibrations, and mechanical misuse.

The BMS design can be of *centralised* or *modular* character. A centralised BMS is equipped with a single control unit connected to the battery cells through a multitude of wires, and this control unit is used to measure, estimate, and predict the operation possibilities for the coming time period. Modular designs, on the other hand, rely on a number of separate controllers, each handling a certain number of cells, with communication between the controllers and with the vehicle control units. Other types of modular design exist, and a common design is the *master–slave* BMS, which can be seen as a combination of a centralised and a modular design using only one control unit for communication with other units onboard.

6.1.1   Charge and discharge control and methods

The BMS must control the charge and discharge levels and make decisions about accepting (or not) brake energy, delivering the requested power output, or using external charging in order to maintain the voltage limits. Maintaining cell voltage within its operating limits, which usually compensates for temperature and current levels, is needed in terms both of safety and durability. Monitoring voltage, currents, and temperature must be performed with acceptable response time in order to estimate and predict the state functions in real time. Given the battery states and the temperature, maximum charge and discharge currents can be predicted to eliminate or limit the possibility of reaching the voltage limits in the next time period, Δt. This time period is needed for the control strategies of the vehicle regarding the coming Δt in order to fulfil the power demands. This is especially important for the energy management of an HEV, i.e. how to utilise the energy distribution between the ICE and the battery for the coming Δt in order to optimise the overall energy consumption.

How fast the charging and discharging can proceed depends on reactions within the cell. These reactions are not instantaneous and there are key processes, related to the charge transfer and the mass transport of the active species in the cell, limiting the charge rates of a battery. Most cell reactions have different time constants depending on the chemical constraints. The charge-transfer process at the electrode surface is fast, within seconds, but the mass transport or diffusion is relatively slow (in the range of minutes or hours) and continues until all active species have been transformed. The slow mass-transfer reactions are the reasons why cells can deliver or accept very high-pulsed currents, but much lower continuous currents.

The concentration gradient in the electrolyte between the bulk and the electrolyte interface will also be a time limiting factor for charge and discharge. Mass transport within the electrolyte and in the solid electrode materials influences charging/discharging time. These processes are, however, temperature dependent, non-linear, and often not the same during charge and discharge.

If the charging rates are faster than the chemical reactions, they may cause local overcharge conditions, polarisations, or a local temperature rise, which can damage the cell. It may be necessary to introduce rest periods during the charging procedure in order to give the mass transport reactions time to proceed to equilibrium. Moreover, the voltage hysteresis (Section 1.6.1.1) will be pronounced if the charge/discharge processes are too fast, and energy will be lost.

Another scenario if the charging/discharging rate is too high is that side reactions may take place instead of the main redox reactions, jeopardising durability and safety of the cells. Due to the higher currents, fast-charging procedures will increase the Joule heating (Section 1.4.3) of the cell, thus increasing the rate of the chemical reactions. The redox reactions (as well as the side reactions) are dependent on temperature and therefore the charging times can vary widely from warm to cold environments. The charge acceptance is considerably limited at lower temperatures, resulting in longer charging times.

The limitations of charge or discharge currents can also be set by the hardware. Different switches or fuses can be used in order to limit or interrupt the currents. One drawback of this set-up is the possible interruption in vehicle operation, which may cause unwanted operation situations and therefore the switches need to be regularly maintained and reset. Moreover, look-up tables based on extensive testing can be used in the control system to set the voltage limits, but the charging constraints are dependent on temperature, SOH, and the usage history, and these limitations are often determined by testing the cells and the battery. Not only the control system and the control algorithms used affect these limitations, the charger device must also be able to charge the battery in a proper way at an optimised rate and know how and when to stop the charging/discharging. Incorrect charge procedures can cause severe damage to the cells or, in the worst case, an abuse situation. Therefore, knowledge and ability to detect when the cells are fully charged and when to terminate the charging before any severe situation appears are essential.

6.1.1.1   Constant current charging and constant voltage charging methods

The simplest of all charging methods is applying a constant, often low, current. The charging rate is normally set to a fixed value depending on the maximum rated capacity of the cell. The drawback is the risk of overcharging the cell and therefore this method must be equipped with robust charging termination control; the two common approaches are voltage control and temperature control.

The *voltage control method* is suitable for NiMH batteries due to their behaviour close to fully charged state (Section 2.2.1). The NiMH technology will therefore be used to illustrate this charging method, which is based on accurate control of voltage changes, $\Delta V/\Delta t$, across the battery over time. As the voltage increases and the cell reaches 100% SOC, the cell voltage drops slightly. This drop can be monitored to interrupt charging. At slow charging rates, however, the voltage drop can be very small or even absent.

The *temperature control method* on the other hand measures the temperature changes ($\Delta T/\Delta t$) during the charging process. As the charging process is continuous and when the cell reaches 100% SOC, most of the charging energy is converted to heat; a change in cell temperature can be detected using a temperature sensor. Figure 6.2 illustrates the two control methods.

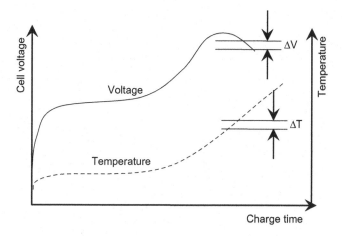

Figure 6.2 Comparison of charge control methods based on voltage changes ($\Delta V/\Delta t$) and temperature changes ($\Delta T/\Delta t$).

The *constant voltage charging method* allows a maximum current to be fed to the battery until the voltage reaches a predefined limit, usually the same as the upper voltage cut-off limit set by the cell manufacturer. This method allows higher currents to be used, thus being more suited for fast charging than the constant current method. The charging procedure can also be divided into different stages, often depending on the discharge profile, before the upper cut-off voltage is reached.

6.1.1.2   The constant current–constant voltage charging method

A third charging method is the *constant current–constant voltage method*, which is a combination of the two previous methods. This method can be used for fast charging and still avoid cell damaging conditions. The charging device must be able to monitor and handle both current and voltage with high accuracy.

The charging procedure can be divided into different phases. Phase I refers to the maximum current flow time period. When the voltage limit has been reached, the current drops gradually to a minimum level, Phase II. In this phase, the battery is charged by the voltage-controlled procedure and the current declines to a minimum level as the voltage approaches a fully charged state. Lastly, Phase III refers to a period of low current flow in the cell, which compensates for the normal self-discharge occurring in all cells. Phase III also allows the chemical reactions to reach a long-lasting equilibrium. A low current is applied until a predefined minimum current level is reached and the charging current is interrupted. Increasing the charging current will not necessarily shorten the charging time. Although the battery reaches the upper cut-off limit faster, the saturation charge time will take longer, accordingly; Phase I will be short, and Phase II will last longer. The duration of Phase II is arbitrary. Figure 6.3 illustrates this charging method and its corresponding phases.

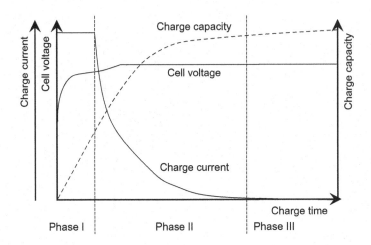

Figure 6.3 Constant current–constant voltage charging method.

By monitoring the voltage and the temperature of the cells, the voltage and the current levels can be adjusted further in order to optimise charging efficiency and time, and extend life, all depending on battery technology and cell conditions.

6.1.2   Thermal control and management

As cell performance, and thereby battery performance, is highly affected by temperature, similarly thermal control and management, in addition to the thermal system (Section 5.2.3.1), also play a key role. Usually, battery performance is specified for a temperature range of *ca.* +20 °C to +40 °C. The temperature of the battery must be held within the limits to optimise performance and life. The performance can differ substantially if the operational temperature is higher or lower. Not only the actual temperature of each cell is important, but also the temperature differences among the cells, which should be held at a minimum. Temperature gradients are likely to be observed within a battery, and consequently the position of the cells within the battery results in different performance and ageing rates. A well-designed battery can minimise the temperature gradients, but it is impossible to eliminate them. Also, the cell balancing is affected by temperature gradients.

The reason why the cells have a better performance at elevated temperatures is due to the rate at which the cell reactions proceed. The rates will increase by temperature, allowing higher power delivery from the battery at elevated temperatures, as well as improving mass transport within the cell, thus reducing impedance and consequently increasing cell capacity.

A drawback with temperatures outside the optimal range is irreversible side reactions taking place, causing increased ageing rates. Therefore, an upper temperature limit is set for optimal battery operation and durability. The same arguments are valid for low temperatures, where the performance can drastically decrease and low temperature operation can cause severe damage to the battery (e.g. lithium plating, Section 3.1.3).

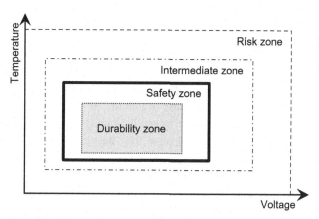

Figure 6.4 Schematic voltage and temperature safety ranges.

Temperature monitoring and management are directly connected to battery safety, and safety can be ensured at many levels. At vehicle level, the mechanical integration is essential to minimise potential abuse and cabin intrusion in, for example, crash situations. At battery and cell levels, there are, in addition, several safety issues to consider in battery design and usage. Keeping the battery, i.e. the cells, within voltage and current limits is one safety level and monitoring cell temperature and temperature gradients within the battery is another. The thermal management should preferably maintain the temperature in a pre-defined region and detect unexpected temperature rises. Figure 6.4 illustrates the temperature and voltage relationships in terms of safety ranges.

Any battery operation will generate heat due to the nature of the cell reactions, as well as the heat generated from losses caused by the resistance within the cell during charge/discharge processes. There are mainly three fundamental sources of heat generation related to losses within the cell: activation losses due to interfacial kinetics, concentration losses due to transport of species, and ohmic losses resulting from the movement of charged particles. The former are related to losses due to overpotentials and the latter is often referred to as Joule heating (Section 1.4.3), often causing temperature rises to about 60–80 °C. The heating caused by overpotentials is not constant during charge and discharge and therefore the heat transfer will vary. The battery heat generation is of transient character, and is dependent on the SOC; the heat generation increases towards the end of charging and discharging. Depending on the technology used, the thermal behaviour may differ. NiMH cells, for example, generate heat during the charging process as the fully charged state approaches and the cell temperature can rise dramatically if the charging current is not terminated. If overcharged, a thermal runaway is a possible unwanted consequence.

6.1.2.1   Thermal runaway
Thermal runaway is a very rapid increase in cell temperature associated with, for example, cell venting, blazes/flames, cell rupture, fire, and explosions. The trigger for

thermal runaway may be overcharging or overdischarging, overheating, crushing, mechanical impact, and external and internal short circuits.

The results of a rapid temperature increase can be severe. For example, reactions caused by overcharge conditions are often generating heat in the temperature range of about 60–80 °C. Normally an overcharge protection device will interrupt the charging procedure before any abuse events, but in case of failures the overcharge may proceed. As for Li-ion cells, they could become severely damaged by overcharge, and, depending on the materials used, the electrolyte may decompose to generate gas. If safety devices are incorporated in the cell (e.g. a CID or a PTC, Section 4.4.4), the CID will open and the charging will be interrupted. The PTC function can also set a lowering of the charge current. At temperatures in the range of 120–130 °C, the key component to interrupt charging is the separator. At such temperatures, the separator deforms or partially melts and thus blocks the ion diffusion in the electrolyte thereby terminating the charging. At higher temperatures (180–190 °C), decomposition of the active electrode materials is triggered, and the safety vents of the cells will open. At these temperature levels, no safety arrangement will be useful in preventing further abuse. The likelihood of emitting gas is high, and undoubtedly a thermal runaway will take place.

In case of an overcharge or overdischarge event, the BMS can often detect the failure since these failures are usually caused by defect connections in the charge/discharge circuit or in the battery or vehicle electronic systems. The same is true if the battery is heated by external sources. The BMS must be able to prevent damage to the battery due to abuse or misuse and this could be accomplished by shutting down the battery, reducing charge and discharge currents, or disconnecting and isolating the battery from the rest of the vehicle.

On the other hand, internal short circuits caused by manufacturing defects inside the cells are impossible for any BMS to detect. Cells are extensively tested at the end of the line, before technology and cell selection, battery assembling, etc. Tests performed in order to evaluate the safety of a cell are mainly performed to understand the cell tolerance to a specific trigger. Most risk situations are, however, not preceded by an obvious external abuse, but spontaneously caused by internal short circuits: a slow development within the cells of internal short circuits that mature to the point where they result in thermal runaway. These internal short circuits are often caused by impure particles originating from the manufacturing process and not observable at the time of manufacturing.

The simulation of internal short circuit caused by impurities could be reproduced by introducing defects in a controlled way. The task is challenging, though, due to the fundamentals of the internal short circuit behaviour that develops only under normal operation of the cell. Thus, fresh cells do not necessarily exhibit the conditions that lead to risk situations. Even though the cleanliness of the manufacturing plants is important, impurities can never be completely eliminated. Fortunately, not all of them will lead to abuse situations and thermal runaway.

The formation of Li-dendrites can also cause an internal short circuit mainly due to non-optimal charging conditions and imbalances in the cell. In the same way, metallic dendrites can be formed due to impurities of, for example, Ni and Fe and these dendrites grow during cycling, as well as during storage. The standard potential for Ni/Ni^{2+} is 2.9 V vs. Li/Li^{+}, implying that Ni will dissolve at typical potentials of the positive electrode and be plated at the negative electrode. Similar behaviours are expected for Fe; the standard potential for Fe/Fe^{2+} is 2.6 V vs. Li/Li^{+}.

These reactions will proceed during storage, which explains how the impurities grow over time. The growth rate depends on several factors: e.g. the size and shape of the particles, the rate capability of the cell, separator permeation and thermal properties, the charge/discharge history of the cell, and the temperature. Migration of particles from one electrode to the other may create holes in the separator thus enabling direct contact between the electrodes. Local melting of the separator caused by hot spots will also increase the possibility of direct contact between the electrodes. By using layered separators, where the different layers have different thermal properties, the risk of an internal short circuit can be eliminated or at least drastically reduced.

During an internal short circuit, the electrochemical energy stored in the battery can be released as heat. Since the cells contain material that can undergo exothermal combustion decomposition reactions (e.g. solvents of the electrolyte and carbon-based electrodes), the heat release of the cell can be much higher than the electrochemically stored energy, which must be considered in the thermal and safety control of the battery. The temperature rise at an internal short circuit is in the magnitude of several hundred degrees per second, which might be impossible for any cooling system to handle. How the thermal runaway proceeds or not, can also be the result of cell design and capacity. Cells of higher capacity and lower impedance will result in more severe conditions than a cell of lower capacity and higher impedance.

It is important to note that not all impurities will lead to internal short circuits and not all internal short circuits will result in thermal runaway. The internal short circuits result in local heat generations, *hot-spots*, and whether or not this heat will trigger decomposition and combustion reactions depends on the temperature and the corresponding temperature increase rate, as well as the nature of the reactions; the temperature must exceed a threshold in order to initiate the combustion reactions. Depending on the cell (i.e. chemistry, capacity, size, and format), different threshold values must be transcended in order to trigger thermal runaway. Often the assumption is made that positive electrode materials with a higher on-set temperature will automatically make a cell safer or eliminate safety incidents, but this is not always true. As for Li-ion cells, the negative electrode decomposition is generally the lower temperature trigger for initiating thermal runaway subsequent to the internal heating generated by an internal short circuit, and this has to be considered as well. The SEI layer growth rate will enhance with increased temperatures, and decomposition of the layer may start at temperatures even below 100 °C, depending on the lithium content in the graphite host structure.

A thermal runaway is often caused by an abnormal temperature increase in one or a small number of cells. This heating of the cell can be a result of several external and internal factors, e.g. external heat sources like a fire in other parts of the vehicle or outside the vehicle, or an external short circuit in the electrical system as a result of, for instance, a collision. A temperature increase may also be caused by using significantly

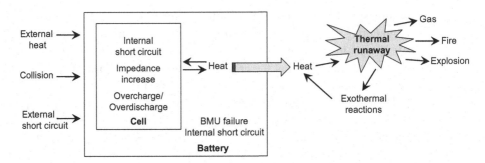

Figure 6.5 A possible thermal runaway sequence in a battery.

high currents or by malfunctions in the BMS, which can cause overcharge or over-discharge. The fact that internal short circuits cause temperature increase is mainly a result of the impurities as described above, or the result of a crash or a collision. The heating will trigger decomposition reactions, which often are of exothermal character, and even more heat will be generated. The consequences will be, for example, gassing, fire, or explosion. A possible course of a thermal runaway is illustrated in Figure 6.5.

As indicated, the thermal runaway in one cell can cascade and trigger other cells to develop thermal runaway, which could, at worst, result in a multi-cell explosion. Moreover, the temperature increase in a cell could be enhanced by the ageing conditions of the cell, which may trigger other decomposition reactions or affect the reaction rates to be more reactive. An aged cell is also more sensitive to overcharge and overdischarge conditions, indicating that the control strategies and control algorithms securing battery performance should preferably be adapted to the age of the cell.

6.1.3   Battery monitoring

There are other safety issues for a battery in an electric vehicle than those related to chemical reactions or temperature. These safety issues are often related to mechanical or electrical faults linked to the electrical system or internally in the battery, e.g. malfunctions in fuses, electrical components used for cell balancing or control, or other faults related to the hardware of the battery. The battery may also be subjected to malfunctions in other parts of the vehicle, e.g. the electric drive system or the control systems, including the BMS. These mechanical and electrical safety issues may arise from vibrations, crashes, external fires, and short circuit protections, etc. If the cells are internally short circuited by mechanical abuse such as crashing or impacts, this should have been detected during the battery and vehicle development phase, resulting in enhanced and more tolerant battery enclosures, as well as a better battery placement in the vehicle, even if it is impossible to design a battery completely tolerant for all impact and crash situations.

Even if temperature is the main safety issue for the BMS to handle, there is always a risk of other cell and battery failures, not due to thermal issues. If a cell is short circuited or goes open circuit in some way, the battery would basically be unusable.

Most probably, the BMS will automatically detect the specific defect cell(s). If the battery has a modular design, it is possible to separate the failing module by independently bypassing and disconnecting the failed module. This safety function, often referred to as the *limp home mode*, allows the vehicle to run using reduced battery capacity. Depending on the modularity, the battery will lose more or less capacity, but ideally only the damaged cell would be disconnected. This set-up can be quite complex, but managed in the right way only the capacity related to the damaged cell will be lost.

The monitoring of the battery is related to the mechanical, electrical, thermal, and chemical character of the battery. Before cell selection and assembly, several safety tests are preferably performed: e.g. forced charge/discharge, external short circuit and heating, vibration and shock tests. Even if different tests are performed, there is no guarantee that the cells, or the battery, are safe. As mentioned previously, it is difficult to perform tests indicating the absence of cell impurities since they might grow during cycling and storage of the cells, and internal short circuits cannot be observed and limited by safety electronics or by the BMS.

6.2   State functions

In order to operate an electric vehicle, the BMS must inform about the power and energy status of the battery, and will rely on information provided by the state functions *state of charge* (SOC), *state of health* (SOH), and *state of function* (SOF). Knowledge about these functions is required in order to optimise energy and power usage, secure performance, and ensure a long battery life. The prediction of dynamic battery behaviour is a critical issue. Accurate estimations of the battery status are desirable and the exact status at any moment is of critical importance to the BMS.

Many factors that affect the operational parameters (capacity, energy, and power output) are strongly correlated and depend on the battery technology used. The influence of each factor is usually stronger during extreme operating conditions. Most of the underlying parameters of the state functions are not directly measurable and must be derived from other measurements, through measurements of voltage, current, and temperature. Estimations and predictions are therefore required to define the status of the battery in any situation. Reliable and robust deployment of the vehicle, and thereby the battery, is a challenge because of the need to assemble numerous cells into a battery able to meet the power requirements, while still maintaining cell performance as individually equal as possible. Data collected from vehicles in usage are often compared to data obtained from laboratory tests. Most methods, however, are based on off-line measurements, and in vehicle applications on-line measurements are required.

In the following sections, the main state functions, SOC, SOH, and SOF, will be described. These functions operate on different time scales; the power capability, i.e. SOF, must be communicated to the vehicle control unit on a (milli)second basis, SOC is valid on the minute scale, and finally SOH on a monthly or yearly basis.

6.2.1   State of charge

The state of charge is one of the most important functions in a battery for electric vehicles, and it is defined as the ratio of available capacity ($Q(t)$) and the maximum possible charge that can be stored in a battery, the nominal capacity (Q_n). Usually it is presented as a percentage:

$$SOC(t) = \frac{Q(t)}{Q_N} \times 100 \tag{6.1}$$

A fully charged battery thus has a SOC of 100% and a fully discharged battery has 0% SOC. The rated capacity or the capacity at BOL is often used as the reference point. The SOC lays the foundation for other control parameters in the BMS, and is the key to proper vehicle control in order to secure power responses due to changes in operation conditions. Moreover, by proper control of the SOC, it is possible to facilitate reliable and efficient utilisation of the battery throughout the battery's life.

In an electric vehicle, the most useful reference point, however, would be the actual capacity at a specific time of operation in order to estimate SOC as accurately as possible. This would result in a control system that has to consider the ageing effects of the battery at every control step. The factors mostly affecting the SOC on a short time basis are temperature and C-rates due to the reduction of effective capacity available. Since the capacity of the cells will decrease as a function of ageing, consequently the SOC ranges will vary over battery life since it is a path-dependent quantity.

There is, however, no simple method to accurately estimate SOC. In an EV application, the SOC is used to determine the remaining driving range, and a high level of accuracy is needed to keep the driver informed about the remaining capacity; the accuracy must be comparable to that of fuel gauges of diesel or gasoline vehicles.

On the other hand, in an HEV application, SOC changes more quickly than in an EV, and in many cases with high magnitudes, due to the acceleration and regenerative braking conditions. The SOC level is also an important control parameter in the overall energy management strategies of the vehicle. The overall performance and fuel consumption will be related to the control strategy and to the interactions between the two power sources. The accuracy must therefore be high in order to improve total fuel efficiency, and secure the requested power performance.

The SOC level is directly related to the properties of the active electrode materials, and different active materials behave differently. In the case of a Li-ion cell, 100% SOC refers to a state where all cyclable lithium ions are inserted in the negative electrode material, and at 0% SOC all cyclable lithium is found within the positive electrode. In the SOC estimation procedures, it is important to distinguish between the lithium content in an electrode and the cyclable lithium content; the two can vary. Cell voltage, and consequently battery voltage, also depend on the SOC level, and decrease with decreasing SOC according to the shape of the discharge profile. The underlying fundamentals of the SOC are thermodynamic, electrochemical, and the material constraints of the selected battery technology and chemistry.

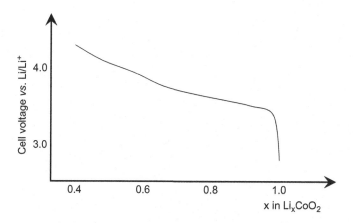

Figure 6.6 Discharge profile for Li_xCoO_2 schematically illustrating the relationship between lithium content and SOC.

The cell voltage can be calibrated to a specific lithium content, and, thus, SOC level, according to the discharge profile. For cells based on Li_xCoO_2, and derivatives thereof (Section 3.2.1), only part of the lithium can be used, i.e. the amount of cyclable lithium is lower than the total lithium content in the active material, due to the stability issues of the lithium-poor structure. In Figure 6.6, the schematic relationship between lithium content and SOC is illustrated. As can be seen, there are more lithium atoms within the structure at 100% SOC, lithium which cannot be cycled.

In parallel to SOC, there is another state function often used related to the available charge of the battery; the *depth of discharge* (DOD) describes how deeply the battery has been discharged, or the amount of charge (in percentage) removed from a fully charged cell. In many cases, the DOD can be seen as 100−SOC.

Depending on the active material of the positive electrode, the relationship between cell voltage and SOC can be straightforward, e.g. in the case of a cell based on Li_xCoO_2 or similar, as shown in Figure 6.6. In the case of other Li-ion cells, it may be more complicated to determine the relationship. The discharge profile for $LiFePO_4$-based cells (Section 3.2.4) is more or less flat, making it more difficult to correlate the voltage to a specific SOC level, and the voltage measurements must be done accurately. A flat discharge profile is often desirable for other parts of an electric vehicle, such as the electric drive system. Therefore, a cell having a flat discharge profile with some distinct, and small, voltage steps would be beneficial. This could, for example, be achieved using a mixture of active electrode materials (Section 3.2.5), and in Figure 6.7 an illustrative discharge profile is shown for an artificial cell having material properties optimised for simplified SOC estimation and constant voltage output.

In vehicle applications, however, it is almost impossible to make accurate measurements of the lithium content in an electrode and thus the SOC must be estimated only based on voltage, current, and temperature measurements, in real time. In order to be as accurate as possible, voltage and temperature should preferably be measured for each individual cell in the battery, but due to the complexity of such a set-up, voltage and

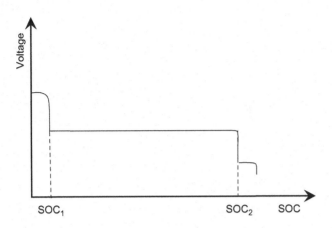

Figure 6.7 Discharge profile for an artificial cell optimised for SOC control and constant voltage output.

temperature are often measured only at a few positions in the battery depending on its size and design. If the battery is made of several modules, the measurements are preferably performed at every module, and, consequently, the SOC of a battery is an average of the SOC per module. Some commonly used methods to estimate SOC in electric vehicle applications will be discussed below.

6.2.1.1 Current counting

Maybe the simplest way to determine SOC would be to fully discharge the battery at a constant rate and count the amount of charge transferred; the capacity is equal to the current multiplied by time. This route is, however, impossible to use in real applications. The current is also not linear over time and therefore the amount of charge should be obtained by integrating the current over time. Knowing the battery capacity, C, and the initial SOC, $SOC(0)$, the present SOC, $SOC(t)$, can be calculated by:

$$SOC(t) = SOC(0) - \frac{1}{C}\int_0^t I(t)dt \qquad (6.2)$$

In vehicle applications, however, the battery is continuously charged and discharged, resulting in an estimated $SOC(t)$ far from the actual value. Therefore, the current counting method requires very accurate current measurements, and the current sensors have to be adapted to the current ranges used in order to minimise the SOC errors. The sampling frequency of the current must also be high in order to capture current peaks associated with acceleration and regenerative braking. Moreover, the Coulombic efficiency of the battery (Section 1.6.5) and the self-discharge (Section 7.1) further limits the use of this charge counting method since both Coulombic efficiency and self-discharge are affected by temperature and SOC.

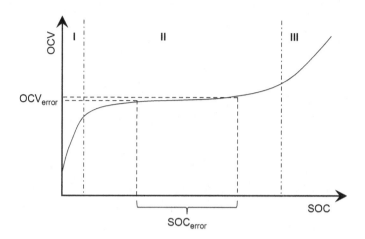

Figure 6.8 OCV as a function of SOC for a cell having a flat discharge profile – emphasising the measurement inaccuracy issue.

6.2.1.2   Voltage look-up tables

Another SOC estimation method is based on voltage. The OCV is a direct method to determine SOC, unusable, however, in real operating conditions. By measuring battery voltage and comparing it with a look-up table or a diagram, it is possible to obtain an estimated SOC value. The actual voltage, however, differs from the OCV. Another aspect to consider is voltage hysteresis, which can further increase the voltage discrepancy from the OCV value.

Many battery technologies, however, lack a strong relationship between SOC and OCV, and as for batteries with a flat discharge profile, a small error in the OCV measurement will result in a large SOC error. This is illustrated in Figure 6.8. In region I and III, a large mismeasurement of voltage results in small SOC errors. On the other hand, in region II, a small mismeasurement of voltage results in a large SOC error. The rapid voltage changes in regions I and II can be used to determine the cut-off voltage. In vehicle applications, however, a more accurate measure is preferred.

6.2.1.3   Battery models

Using a battery model as the basis for SOC estimations has been the common approach for vehicle applications. Input parameters are data from current, voltage, and temperature, which must be sampled at a fixed rate sufficiently fast in order to follow the usage conditions, i.e. the often rapid changes between power demands.

The battery model must be detailed yet easily implemented, and able to handle the rapidly changing variables during vehicle operation. The order of the model is therefore an important aspect of SOC estimation. A high-order model often results in dependency between different estimators that are complex to implement in real time. On the other hand, low-order models may provide insufficient accuracy of the estimated SOC value. Therefore, a trade-off between capturing the battery dynamics and estimation accuracy is preferred. The accuracy of the battery model depends on several variables:

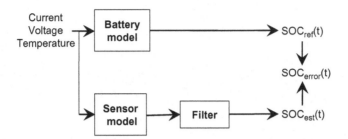

Figure 6.9 Battery SOC estimation procedure including a battery and a sensor model.

geometrical, electrical, and, especially, electrochemical. During model development, these variables must often be determined by extensive cell and battery testing. Profound knowledge of all underlying charge and discharge mechanisms is the key to model trade-off. Perfect models are, however, seldom achieved in representing the actual battery dynamics and errors must be considered in the model-based SOC estimation. In addition, corrections and feedback to the model must often be taken into account. One of the main tasks of the battery model in the SOC estimation process is to define the reference SOC value since the initial value is seldom known.

The SOC battery models are often based on filter functions, e.g. a Kalman filter, to combine the voltage look-up and current counting methods for real time conditions. A Kalman filter is a commonly used method to estimate the state variables of dynamic systems, and it employs a series of measurements observed over time, containing random variations and other inaccuracies. The state at the present moment is estimated via the state and its covariances of the previous moment, and the information gained at the present moment is used to refine the state estimation of the following moment, and so forth. The outcome is estimations of unknown variables that are more precise than those based on a single measurement alone. Due to the non-linearity of the battery dynamics, an extended Kalman filter is used to approximate the non-linearity to a linear response at every moment of estimation.

One uncertainty for all these SOC estimation procedures is the accuracy of the sensors used, and the corresponding sensor noise. The effect of SOC errors on performance prediction also depends on the response time of the model compared to the actual battery. It is important to adjust the models in order to achieve a reasonable time for the model to converge. Using a reference electrode (Section 1.7.4) would be favourable in order to obtain a more accurate voltage determination, but this is not viable in real vehicle applications. There are, however, several other ways to determine SOC estimation accuracy, or SOC errors, e.g. mean errors including standard deviations, maximal errors, or root mean square errors. Most of these methods require accurate battery models in order to give reliable results. How to combine the different measures in an SOC estimation procedure is illustrated in Figure 6.9.

Methods other than Kalman filtering can be applied to improve the accuracy of SOC estimations, such as fuzzy logic and self-learning adaptive methods. Without exact equations and precise data, the fuzzy logic method can give definite estimations of, for example, noisy battery current data and the method is commonly used in automatic control systems. It may, however, be useful to employ several SOC estimation methods to improve the accuracy by weighted estimates.

6.2.1.4 Balancing

Preferably, all the individual cells in a battery have the same characteristics regarding capacity, SOC, temperature, etc. During operation and storage, the cells will end up differently. Uniform cell characteristics are preferred in order to secure performance within the safety limits of the battery. The cells can also differ in capacity due to variations in the manufacturing processes, and experience uneven ambient conditions in the battery due to the electrical and thermal environment.

Cell imbalance within a battery may also be the result of external sources, like circuit configurations and ambient temperatures. Cells subjected to rapid charging/discharging conditions, e.g. in HEVs, are more affected by the cell-to-cell variations since the imbalance will be accelerated by the rapid changes in the current directions. Even if the differences are negligible at BOL, they may severely affect the performance characteristics and cause degradation during battery life.

Furthermore, cell imbalance can lead to failures, abuse situations, as well as shortened life. The potential failure rate is higher than can be expected by just multiplying the cell failure rate by the number of cells. The major reason is that interactions between the cells in the battery are essential. Furthermore, the production process can result in small, but not negligible differences, leading to uneven utilisation of the cells and, consequently, inhomogeneous ageing. Therefore, battery design should preferably be based on cells being as equal as possible.

The cells connected in series and parallel in a battery will behave differently if there are deviations between the individual cell characteristics. Cells connected in series will experience the same current and, in theory, these cells would balance themselves if they were identical and had the same environmental conditions, such as temperature. Cells connected in parallel are automatically balanced due to their equal output voltage. The variations in cell impedance among the cells will, however, affect current distribution, and imbalance may occur for parallel-connected cells if, for example, one of the cells is short circuited, causing the other cells to discharge.

It is also possible that individual cells could be overcharged/overdischarged, resulting in premature cell failures. One cell may reach the cut-off voltage before the others, not because it is fully charged or discharged, but because its internal impedance is higher than the other cells, i.e. the cells are in practice subjected to different C-rates. The cell will subsequently have a deviant SOC and thus be subjected to greater stress, resulting in accelerated ageing. Usually the degradation of the cells is observed as increased cell impedance and resistance, and the imbalance will be amplified. During operation, if there are degradation variations among the cells, there is a risk that these cells will reach the fully charged/discharged limits before the less aged cells. If there are cell failures in a battery, it is tempting to replace the failed cell by a fresh cell. This will, however, most likely increase the imbalance among the cells even further since the impedance of the fresh cell is quite different from the other cells, and failure will undoubtedly occur again.

For all of the reasons above, a cell balancing system is required to equalise the cells as much as possible in order to secure uniform performance, durability, and safety of the battery. Various methods to balance the cells have been developed, trying to equalise the stress on the cells and thereby prevent long-term degradation differences rather than small short-term deviations. All balancing methods depend on the ability to determine the SOC of individual cells. There are two main categories of cell balancing methods: *passive* and *active*. Passive cell balancing tries to equalise the voltage of all cells by consuming the charge/energy of the cells, while active balancing redistributes charge/ energy between the cells; both methods are described below.

Passive balancing

The passive balancing removes excess charge (i.e. energy) from the fully charged cells in order to achieve cells with as equal charge levels as possible. This can be achieved in many ways, e.g. by switching off the charging/discharging current when the first cell has reached the cut-off voltage. This procedure will, however, contribute to incomplete utilisation of the battery; either the battery will be only partly charged or capacity will not be fully utilised. This, in turn, reduces battery efficiency and may limit energy utilisation of the electric vehicle.

Other passive balancing methods interrupt the charging current as the first cell reaches cut-off voltage and then discharges that particular cell until all cells reach the same voltage or charge level. Still other methods continuously charge all the cells until they are fully charged, but the voltage applied to the individual cells is limited by using, for example, resistor elements, until all cells are equally charged.

Passive balancing based on shunting resistors is a simple equalisation concept. Once a cell reaches a pre-set voltage, it is bypassed until all the cells reach the targeted voltage. The shunting resistor elements can be either in fixed or switched mode, respectively, as illustrated in Figure 6.10.

The fixed shunt resistor method continuously bypasses cells by adjusting their cut-off voltages. This method can only be used for cells tolerant to overcharge conditions, e.g. Pb-acid and Ni-based technologies. On the other hand, the switched shunting resistor method controls the bypassing by switching relays. The relays control the cut-off voltage, either at the battery level or by monitoring the individual cell. Methods based on switched shunting resistors are reliable with respect to avoiding overcharging and can therefore be used also for Li-ion cells. As shunted energy is dissipated as heat, the thermal management must be designed to eliminate large temperature differences within the battery caused by the cell balancing.

Active balancing

The active balancing methods remove charge from cells of high SOC and deliver it to cells of low SOC, according to the 'Robin Hood principle': to steal from the rich and give to the poor. Since it is impractical to charge the cells individually, the charge must be applied sequentially. There are several active balancing methods to make the equalisation process fast enough. One example is to interrupt the charging/discharging procedure as a cell reaches the cut-off voltage and then continue to charge/discharge the

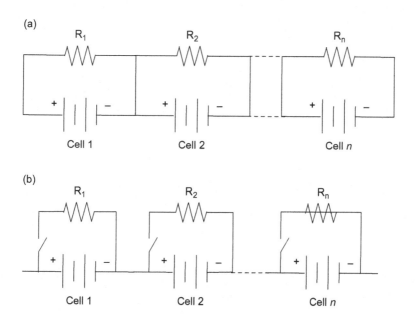

Figure 6.10 Passive shunting resistor balancing topologies (a) fixed resistors, and (b) switched resistors.

other cells until they have all reached the predetermined voltage level. This will maximise the capacity of the cells and consequently maximise the capacity of the battery. Different topologies are used according to the active elements, e.g. capacitor and/or inductive components, as well as controlled switches or converters.

Capacitive balancing methods use capacitors to shuttle the charge among the cells in order to find balance. The methods are therefore sometimes called 'charge shuttling' methods. The basic principle is to transfer energy from a cell with a higher voltage, via the shuttle capacitor, into a cell with a lower voltage, resulting in a redistribution of the energy. The efficiency is reduced, though, if there are large voltage differences among the cells. The shuttle capacitor can be designed in different topologies, e.g. basic switched, single switched, or double-tiered switched. The basic switched capacitor method is illustrated in Figure 6.11.

Inductive charge shuttling balancing methods use a transformer having its primary winding connected across the battery and its secondary winding switched across the individual cells. It levels the capacity just like the capacitive balancing methods, but by pulses of energy when requested, without using the small cell voltage differences.

6.2.1.5   Estimation for capacitors

The SOC estimation for an electrochemical double-layer capacitor (Section 2.4) is more or less straightforward as the voltage across the terminals ideally determines the SOC, and a typical discharge profile is given in Figure 6.12.

The maximum charge, Q_{max}, stored in a fully charged capacitor is related to the capacitance, C, given in farads, and the maximum voltage, V_{max}:

$$Q_{max} = CV_{max} \qquad (6.3)$$

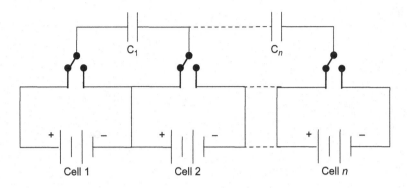

Figure 6.11 Example of active balancing: the basic switched capacitor method.

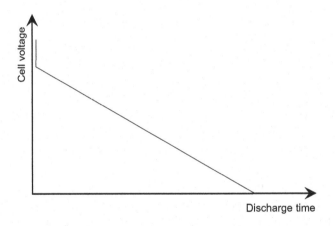

Figure 6.12 Typical discharge profile of an electrochemical double-layer capacitor.

The corresponding SOC is therefore:

$$SOC = \frac{Q(t)}{Q_{max}} = \frac{CV(t)}{CV_{max}} = \frac{1}{CV_{max}} \int_0^t I(t)dt \qquad (6.4)$$

As the voltage of a capacitor increases linearly with charge, the storable energy in a capacitor is proportional to the square of the voltage, as described in equation (2.2), which can be further developed according to:

$$Energy = \int_0^Q V dq = \int_0^Q \frac{q}{C} dq = \frac{Q^2}{2C} = \frac{Q_{max}^2 SOC^2}{2C} \qquad (6.5)$$

Thus, the stored energy in a capacitor is proportional to the square of the SOC, and an accurate estimation is much preferred in order to enable use of the capacitor in an optimal way in an electric vehicle.

The relationships so far are valid for an ideal capacitor and at OCV conditions, preferably at relaxed conditions. The terminal voltage is often inaccurate as it excludes the physical state of the capacitor, and often non-linear behaviour is observed due to electrical and chemical phenomena, especially for asymmetric capacitors. Therefore, model-based SOC estimation methods are used similarly to the methods used for electrochemical battery cells. In an EDLC, the voltage changes quickly during charging/discharging processes, which is why the voltage sensors used must be of high accuracy in order to minimise errors, and thereby maintain reliability and control of the vehicle.

6.2.2   State of health

If SOC is the indicator of remaining battery charge, the *state of health* (SOH) is the indicator of remaining full battery capacity, compared to capacity at BOL. The ageing processes lowering SOH commence once the cells are assembled. As the battery ages, the SOH will decrease until the battery can no longer fulfil the requested performance (i.e. EOL). Therefore, the reliability of the SOH prediction is crucial to the control of an electric vehicle.

Predicting the remaining life is a challenge due to the complexity of performance degradation. Therefore, it is essential to monitor performance concerning long-term changes in the battery, since the SOH is path dependent. SOH is not a physical quantity, but it depends on and can be represented by several physical parameters: e.g. the number of charge–discharge cycles, capacity and power fade, increase in impedance or internal resistance. These parameters affect different degradation mechanisms within the cell, mechanisms dependent on cell chemistry and design, manufacturing processes, and usage.

Batteries age in complex ways, depending on the mechanisms involved, and in a variety of combinations, resulting in path-dependent behaviour. SOH is affected by the degradation mechanisms occurring within the cell during the entire life of the cell and how the cell reactions evolve over time. The ageing factors can arise from vehicle usage, and battery and cell design, and main ageing factors are temperature, SOC range, energy, and power conditions, as well as time, both usage and calendar time. The ageing rate depends on operational conditions encountered over battery life. Changes in the usage conditions (i.e. the ageing conditions) may accelerate or decelerate the degradation mechanisms, and even initiate new ones. Path dependency is therefore a key element in predicting the remaining battery life, but can be complicated to follow. Therefore, it is important to regularly track the usage history of the battery.

Battery life is often divided into calendar life and cycle life; calendar life refers to the number of years a battery can be used in a specific application, and cycle life is estimated for specific performance criteria, and refers to the number of cycles the battery can execute until the performance criteria are no longer met. Cycle life depends on parameters like power and energy, temperature, C-rate, and SOC range.

The ageing rate is often defined as the changes in SOH as a function of time, where the time scale for SOH is normally months or years:

$$Ageing\ rate\ =\ \frac{\Delta SOH}{\Delta t} \qquad\qquad (6.6)$$

In power-demanding applications, like HEVs, resistance and impedance development have a strong impact on power performance. In EVs, the energy requirements are dominant and the ability of the battery to maintain the voltage is important. Consequently, at the vehicle level the SOH can be detected as a decrease in performance, mainly as power fade or reduced all-electric driving range.

In HEV applications, both usable capacity and available power can decrease over time, thus limiting the performance of the battery and, consequently, the energy efficiency of the complete vehicle. The driver will notice the degradation through, for example, increased consumption of fuels for the ICE. In the case of an EV application, the driver will observe a decrease in the driving range as a consequence of cell ageing. These observables are due to different losses in the cells – losses of energy, power, and capacity – all related to the SOH of the battery, and often within the cells.

Cell degradation and ageing are often a combination of these losses, which simultaneously affect each other, and most of the losses are related to irreversible chemical reactions within the cell. The loss of energy is related to the ability to maintain the voltage level of the cell and is observed as a decrease in capacity. The power losses are due to resistance increases resulting in a reduced ability to deliver (or accept) charge at requested rates. Even if the power fade affects the practical performance of the battery, it is not a suitable measure to predict ageing or SOH, since it depends on several factors, such as SOC range, and battery and ambient temperatures. The SOC range available will be limited as the cells age and therefore the SOH will in turn also affect the SOC. Moreover, cell-to-cell variations in the battery may influence the power capability.

A number of different models have been developed to estimate and predict SOH and the long-term behaviour of cells. The models applicable for SOH estimation are generally classified as: (i) physical, (ii) equivalent circuit, (iii) empirical, or (iv) combinations thereof. The challenge of all models is to find time-resolved quantifications of degradation and to connect microscopic causes and macroscopic observations; the input for reliable battery diagnosis.

6.2.2.1   Physical models

In the *physical models*, ageing and degradation are determined by fundamental physical and chemical relationships. These models tend to be very detailed and complex, but provide results of high accuracy. Electrochemical fundamentals like the Butler–Volmer equation (equation (1.14)) and porous electrode theories are basic conditions for the models. Usually the degradation phenomena related to the material properties and the intrinsic material properties are taken into account, and *ab initio* calculations and molecular dynamics are often used to establish understanding. Depending on the nature of the materials, the different phenomena occurring during lithium insertion/extraction reactions, e.g. phase transitions processes, may be observed and described. The models require fundamental understanding of the underlying reactions occurring in the cell, during cell usage, as well as during storage. The challenge for these types of models is to link the results to macroscopic observations, since the aim of the models is to describe in detail the degradation mechanisms occurring at the atomic level. Thus, estimating the direct impact of the results on battery performance is a challenging task. Moreover, though viable, it is difficult to use the physical models for diagnosis due to the challenge of including path dependency.

6.2.2.2   Equivalent circuit models

Equivalent circuit models are based on different electrical components, e.g. resistances, capacitances, inductances, to model the behaviour of the cell rather than the electrochemical processes taking place. Commonly, battery behaviour is modularised into equivalent circuits and different techniques are used to estimate the model parameters required, e.g. voltage and current responses, impedance and resistance increases. The parameter identification can be directly performed from measurements or from physical-based models. Even though it requires large data sets, usually obtained from time-consuming tests, it may be a challenge to convert actual usage conditions into laboratory test procedures. To their advantage, these models are easily connected to other parts of the electrical system or the control system of the vehicle. On the other hand, they often suffer from lower accuracy.

6.2.2.3   Empirical models

The *empirical models* often become specific for the given battery technology of a particular electric vehicle, and applying the model to other batteries is often complicated. The degradation estimations are often used without considering calendar and cycling ageing constraints, thus looking at ageing in more general terms. Usually the different ageing factors (temperature, current, voltage, etc.) are correlated to the capacity fade in order to find a relationship for capacity degradation over time. The duty cycle data normally used as inputs for the models are often divided into parts of comparable performance characteristics to find correlations between operational parameters and battery ageing. The main drawback of empirical models is the insufficient prediction of electrochemical processes contributing to capacity fade and impedance rise. Furthermore, the impact of each ageing factor must often be examined independently, resulting in large test matrixes. Depending on the test procedures used, the model may result in considerable errors if, for example, accelerated ageing test procedures are used.

6.2.2.4   Other models

Other simplified models to determine the remaining life of a battery are the *cycle counting* and the *energy throughput counting* models. The cycle counting model, quite logically, counts the number of cycles performed. The ageing factor used is only the total SOC range used, ΔSOC, and the model is based on the assumption that the life of the battery is determined by ΔSOC. Thus, the greater the range, the more life will be consumed and fewer cycles can be performed before EOL is reached, as is schematically illustrated in Figure 6.13.

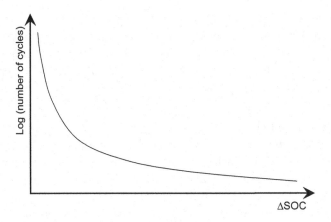

Figure 6.13 Schematic representation of the number of cycles as a function of ΔSOC.

In order to make this model more applicable, temperature and/or C-rate may be considered. This model is mostly used in the early stage of the battery design process, but it is generally not recommended if accurate ageing estimations are required since material degradation effects are excluded.

The energy throughput counting model is based on the same principles. The ageing factor considered is the energy throughput, which is quantified by simply counting the amount of energy (or capacity) passing through the battery. The actual energy throughput used up to a given time is compared to the total energy throughput the battery can withstand until EOL is reached. The remaining life can be calculated given the same ageing rate per used energy unit. The principal basis of the model is the assumption that battery life is reduced proportionally to the energy throughput, and all other ageing factors are excluded. More extended variants of the model can, however, include weighted ageing factors such as temperature, ΔSOC, and C-rate. The dependability of these ageing factors is often not considered, and similar to the cycle counting model, the ageing and degradation of cell materials are excluded, factors which may rapidly reduce battery life.

Taking the above into consideration, it is obvious that a trade-off must be made between complexity and accuracy, and thus combined models are often needed. There are several test methods commonly used to estimate and predict the degradation of the cells and the corresponding losses. These methods will be discussed in conjunction to the discussion about the usage and degradation of Li-ion batteries in Section 7.2.

6.2.3   State of function

One key issue of the control system is to determine the capability of the battery to deliver and accept charge at a given moment and is used to obtain the *state of function* (SOF), which refers to the capability to fulfil demand, often power requirements, at a specific time. This is sometimes referred to as the *state of power* (SOP), when the only parameter included is the power capability of the battery at a given moment. In the SOF, however, temperature, safety, and SOH status are also taken into account. Traffic situations and weather data may also be incorporated, via, for example, GPS data.

Through the SOF, the battery BMS communicates with the vehicle control unit. The SOF can be seen as a unifying term indicating the actual function of the battery, including SOC, SOH, and the temperature and surveillance conditions at a given time. The power capability of the battery and the prediction whether the battery can provide the necessary power are the main issues of the SOF; both short-term and long-term predictions are required. At low SOC levels, there are limitations in discharging power and at high SOC levels the capability of the charging power is limited. The remaining capacity in the battery may not be accessible depending on the cell impedance conditions, e.g. after a rest period the impedance of the cell may decrease, and despite the same operating conditions the SOF can be quite different before and after the rest. The useful capacity available also depends on the C-rates, and here both the Peukert and Ragone plots (Sections 1.4.3–1.4.4) are important tools for optimal control strategies. The dynamic behaviour of the vehicle, and corresponding dynamics of the battery, varies from microseconds to minutes, hours, and years. Therefore, prediction capability is one of the issues concerning optimal performance and durability of the battery. SOF may be of vital importance for energy management in HEVs where the decisions must be made more or less directly if the requested power demands are to be handled by the electrical system or the ICE.

The SOF can be estimated in several ways and is strongly dependent on the SOC, SOH, and temperature, and the non-linear nature of the battery impedance results in a non-linear prediction of the SOF. The short-term power capability is needed in order to fulfil the usage condition requirements. This does not mean that the remaining energy is accessible to the particular application, i.e. the power request may be greater than the capacity can handle due to SOC, temperature, and SOH prerequisites, and is communicated to the vehicle control unit by the SOF.

The temperature of the battery is essential for the SOF estimation, and an even temperature distribution is required inside the battery. Moreover, the capacity imbalance for series-connected cells, i.e. the cell with lowest capacity, will determine the SOF for a specific moment, and results in non-optimal usage. The accuracy of the voltage, current, and temperature measurements is essential for the SOF predictions and should fit the requirements of the chemistry, as well as the application. Different control parameters need to be easily monitored in order to detect if a specific value is exceeded, while other parameters need to be given in real time. The charge control often requires accurate measurements instantaneously, while SOH may require less stringent monitoring of SOF estimations due to the different time scales. Safety is an underlying parameter in the SOF, and the charge control is therefore essential to avoid overrunning the safety levels. The key is to use monitoring and control devices matching the application requirements. Since the SOF is significant to the diagnosis of the battery, the set-up of the algorithms behind the SOF should preferably include fault detections.

7   Battery usage and degradation

In the previous chapters, vehicle and battery requirements have been discussed, in connection with the design constraints in terms of cell selection and battery management and control. Now it is time to discuss usage and durability of the battery in electric vehicles. The durability of the components is of key importance in order to design economically and environmentally sustainable vehicles. Many components are designed to last as long as the vehicle itself, which is about five to 15 years depending on the type of vehicle and market. Since the battery is the most costly component in an electric vehicle, it is important to design it optimally for the intended application with respect both to usage and durability. Moreover, the battery is possibly the most sensitive component. In the following sections, ageing and durability of batteries will be discussed, from the role of usage conditions for durability to the details of degradation mechanisms of the active and non-active materials within the cell.

The goal is to provide knowledge of how batteries age and degrade, including the operating parameters affecting the degradation, in order to enable the designer to postpone the EOL as far as possible. To understand the correlation between utilisation and degradation is specifically important for HEVs where energy management can be adjusted in a different way compared to EVs; the HEV has two energy sources to be optimised for battery durability and fuel savings.

Battery cells start to degrade as soon as they are assembled. There are several factors influencing the life of a battery: e.g. temperature, current rates and direction, SOC ranges, mechanical effects like vibrations – all having a negative impact on battery life. The usage history will also have a considerable effect on durability, in the same manner as it influences the state functions. Strong path dependency and a complex matrix have to be understood in order to get a full picture of the degradation and to predict durability and hence the remaining life of the battery. To know the possible degradation mechanisms helps in designing a battery, and a corresponding management system, both for performance and life.

Normally durability of a battery is compared to a battery at BOL conditions. The measure can be number of years, number of cycles, and total energy throughput. Depending on the type of vehicle, different quantities are used. Vehicles relying on the battery as a primary energy source, i.e. EVs and, to some extent, PHEVs, are more sensitive to capacity fade than power fade since the capacity fade is directly related to the all-electric driving range. On the other hand, HEVs, having the battery as a buffer for short high-power pulses and using a minor part of the installed capacity, are more sensitive to power fade than capacity fade. The power fade will mainly be observed by increased fuel consumption.

Battery performance will decrease whether the battery is used or not and, subsequently, storage time affects battery life; an important design parameter for passenger cars, which are mostly parked. Durability or life of the battery is often categorised as *calendar life* and *cycle life*. The former is often determined by number of years and the latter in maximum number of cycles for specific operating conditions.

The calendar life, or calendar ageing, is defined as the irreversible capacity loss during storage, and is mainly affected by storage temperature and SOC level. The calendar ageing appears to follow Arrhenius-like relationships, thus a linear relationship with time may represent the capacity fade and resistance increase. The impact of SOC on calendar ageing does, however, not follow the same relationship, instead it is related to the electrochemical reaction rates due to overpotentials, and the resulting capacity fade and impedance increase are therefore not linear with time. Test procedures to determine the calendar-ageing rate are often specified to hold the cell at a constant charge level, and thereafter the cell is discharged to a predefined SOC level and left in storage mode at a constant temperature. The tests normally continue for longer periods of time (weeks, months, or years) and increasing the temperature may accelerate them. The cells could also be stored without being connected to any device, and in such cases the cells are charged as in the case above, and once they reach the predefined SOC level they are left for a specified period of time, usually weeks or months. This kind of test procedure is focused on shelf-life more than calendar life.

There are two main observations of battery ageing: capacity and power fade, both of which are mainly related to losses of active material and cyclable lithium inside the cell. Capacity and power fade differ due to chemical causes with different origins, causing complex dependency of cell ageing. The performance losses are caused by various mechanisms depending on the chemistry used, and cell design and assembly, and they can be of either chemical or mechanical character. The consequences are primary losses of cyclable lithium, and secondary losses of active electrode material, both resulting in decreased capacity. Side reactions inside the cells will increase cell resistance, and lower the maximum power of the battery. Furthermore, it is important to remember that the same chemistry and cell design may exhibit different performances due to manufacturer and production lot, and therefore specific cells can degrade faster than others.

7.1   Degradation basics and mechanisms

From a vehicle perspective, degradation of the battery cells is normally observed as a decrease in the electrical driving range, fuel savings (for HEV and PHEV), and power performances. All these properties originate at the material level inside the cells: degradation due to ageing of the materials, active, as well as inactive. The performance degradation can be summarised in two categories: chemical and mechanical. The main chemical degradation is observed as decrease in capacity, increase of the self-discharge rate, and decrease in power as a result of increased internal resistance and impedance.

Mechanical degradation effects originate both from inside the cell and from the outside, and are less distinct than the electrical degradation. Some mechanical degradation processes will however lead to loss of electrical contacts, like exfoliation of electrode material from the current collector or corrosion of cell-to-cell connections.

All degradation mechanisms result in three main losses: energy, power, and capacity. The loss of energy is related to the ability to maintain cell voltage, which is often related to changes in the active electrode materials. The power capability of the cell is related to the ability to deliver/accept charge and will be reduced due to the increase in cell resistance and impedance, caused by degradation of the active materials and by side reactions. The loss of capacity is affected by the ability of the materials to preserve charge, and the main cause for capacity loss is the change in balance between the electrodes. This is a result of the reduction of cyclable ions, and the reduction of electrode material and area.

The abilities to maintain both voltage and charge, and to deliver charge are all due to the material properties of the cells and how these are affected by different driving conditions, as well as storage conditions like ambient temperature, charge state, age, and time. It is important to remember that the degradation starts at the very moment of cell assembly and cannot be eliminated, only inhibited.

Knowledge of capacity fade and the power capability of the battery are of key importance not only for battery life, the BMS must be able to accurately predict the SOF. The control parameters to be measured and controlled are related to usage and operational conditions:

- current: charge and discharge rates, especially high currents
- voltage: upper and lower cut-off voltages
- SOC: range and variations
- temperature: maximum and minimum, as well as temperature cycling
- energy throughput

Power decrease and capacity fade can have several root causes and stress factors, as listed above. The most dominant degradation mechanisms in Li-ion cells suitable for electric vehicle applications are related to material degradation of the composite electrodes, and to side reactions occurring at the electrolyte–electrode interfaces. From a material point of view, there are three main attributes related to power decrease and capacity fade originating at the material level inside the cells: changes in active material, changes in inventory of active species, and changes in ohmic and faradic resistances.

The capacity fade of a Li-ion cell can be illustrated by dividing it into different regions according to Figure 7.1.

The capacity loss at the beginning of usage, region I, takes place during a number of cycles. This rapid decrease in capacity can be explained by the formation of the cell and loss of cyclable lithium due to the build-up of an SEI layer on the carbon-based electrode (Section 3.1.1), leading to a declining formation rate. Using a non-carbon-based negative electrode that does not form SEI layers may limit this region.

Most of the battery operation thus takes place in regions II and III. As a result of the SEI formation, a reduced amount of the active material of the negative electrode is

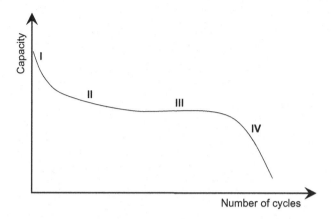

Figure 7.1 The capacity fade of a Li-ion cell divided into regions having different degradation causes.

available for lithium insertion during the charge process of the cell, thus the limiting factor in region II is the negative electrode. The SEI layer may however crack during cycling and new active electrode material is exposed to the electrolyte, resulting in new SEI being formed and consumption of even more cyclable lithium. This further loss of cyclable lithium will result in less lithium ions being inserted in the positive electrode during the discharge of the battery. In region III, the degradation and loss of active material in the positive electrode starts to be accountable, even if the negative electrode is still the dominant factor of the degradation. The loss of active material in the positive electrode will continue and in region IV the amount of active material is less than the amount of cyclable lithium ions, and more and more lithium ions are caught in the negative electrode.

The EOL, normally defined as 80% of the initial capacity, is usually found in region III, preferably as close to region IV as possible. It is difficult to predict the different transitions from one region to another, but region III should be as long as possible, and the EOL should be reached in region IV.

The ageing contribution of the different regions is strongly dependent on the active materials used both for the positive and the negative electrodes, and different degradation processes are dominant in different regions. For example, the sensitiveness of the positive electrode material to overcharge and temperatures will highly affect the degradation rates. The trends presented in Figure 7.1 can thus be seen as very general. The key in designing a battery for long life is to understand the transition from one region to another, and to set up control strategies to prolong region II and III as far as possible or at least to optimise the number of cycles before entering region IV.

The different degradation processes occurring during operation of a cell can be modelled to provide indications of the dominant losses during different parts of the ageing. The degradation mechanisms of loss of cyclable lithium, loss of active electrode material, and a combination thereof, will hence be illustrated by schematic discharge profiles and balancing diagrams. In all examples, the cell comprises a negative

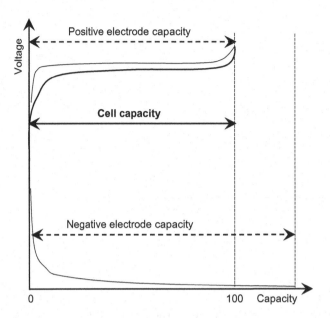

Figure 7.2 Cell balancing and capacity, incl. discharge profiles, for a cell and the corresponding electrodes for a Li-ion cell having a larger negative electrode capacity than positive electrode capacity (i.e. the capacity is limited by the positive electrode).

electrode of graphite as the active material, and an excess of negative electrode capacity will be used, i.e. the cell capacity will be limited by the positive electrode. Figure 7.2 illustrates the cell capacity and the electrode balancing, including the discharge profiles, for the positive and the negative electrodes, respectively, and for the corresponding cell.

As the negative electrode capacity is larger than the positive, the latter will limit the charging procedure proceeding towards 100% SOC, and during discharge both electrodes will limit the operation proceeding towards 0% SOC.

All cells in a charged state deteriorate as a result of the chemical reaction proceeding to an equilibrium state. This *self-discharge* proceeds whether the cell is used or not, and is mainly due to battery technology, cell design, and temperature. The self-discharge is normally a temporary loss of capacity and the rate can be determinant, for example, in the case of a vehicle parked for long periods without any usage at all. Different types of battery chemistries and cells have different self-discharge rates, normally measured as capacity retention of storage given in percent decline per time period (usually per month). Self-discharge is not a defect arising from the manufacturing process, even if poor manufacturing processes and control can influence the self-discharge rate negatively due to, for example, impurities and a high internal resistance. In case of stored cells for long periods, the self-discharge can proceed until the cell voltage falls far below the lower cut-off voltage. In such a case, the cell will most likely not recover.

Normally the storage time, i.e. the parking time, of a passenger car is interrupted by charge–discharge cycling during vehicle operation, and thereby the self-discharge is normally negligible. The effect of the cycling during operation will strongly affect battery life and is a cross-dependency of calendar and cycle life. On the other hand, commercial vehicles (e.g. buses and trucks) are used most of the time and the parking time is often non-existent, and the dominant ageing is therefore related to cycling. Cycling life, or cycling ageing, on the other hand, is related to the charge and discharge procedures taking place during the usage of the vehicle, consequently ageing is a result of charge/discharge procedures, SOC range, duty cycle, and temperature conditions. A major part of the following sections is related to the ageing caused by battery usage, i.e. cycle life degradation.

7.1.1   Examples: origins of capacity fade

To illustrate in more detail the origins of capacity fade some examples are provided here to show individual origins, as well as how they can combine.

7.1.1.1   Loss of cyclable lithium

By design, the capacity of the negative electrode is usually larger than the positive electrode in order to compensate for the formation of the protective SEI layer, formed during the first cycle in order to stabilise the electrode/electrolyte interface. During the first charge of the cell, cyclable lithium ions will be lost in the SEI layer, and at the subsequent discharge the positive electrode will have unoccupied sites available for lithium insertion. Even more cyclable lithium ions will be lost if the SEI formation process continues during cycling caused by side reactions. Due to the losses of cyclable lithium ions, the number of available sites in the positive electrode will increase with each cycle, reducing the cell capacity further, i.e. the capacity of the positive electrode is not fully utilised. Figure 7.3 illustrates a charge–discharge cycle where the capacity loss is only caused by the loss of cyclable lithium.

At the beginning, all cyclable lithium ions are found in the positive electrode. Reaching the fully charged state, some lithium ions are lost in the SEI formation, and thereby the negative electrode is not fully filled with lithium. During the subsequent discharging, less cyclable lithium ions will be inserted into the positive electrode than during the first discharged state. The limiting electrode in this example is the positive. The loss of capacity will continue for every charge–discharge cycle, and will be further enhanced by side reactions.

Thus, neither the positive nor the negative electrode is utilised to its full extent and this loss of cyclable lithium can be further illustrated according to Figure 7.4. The positive electrode will continue to limit the cell capacity by reaching its fully charged state before full capacity has been utilised.

7.1.1.2   Loss of active electrode material

During operation, both positive and negative electrodes may undergo irreversible damage resulting in permanent capacity losses. These damages can be results of, for example, impurities and structural changes due to overcharge/overdischarge conditions; the losses could occur independent of capacity decrease due to the loss of cyclable lithium above. In Figure 7.5, the irreversible loss of active material, both at the positive

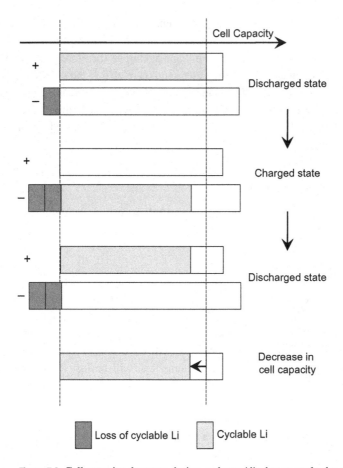

Figure 7.3 Cell capacity decrease during a charge/discharge cycle due to loss of cyclable lithium.

and the negative electrodes, is illustrated, including the loss of cyclable lithium due to side reactions other than the SEI formation.

The capacity decreases due to the increased imbalance between the electrodes, and the corresponding electrode discharge profiles, as well as the cell discharge profile as illustrated in Figure 7.6.

7.1.1.3   Loss of cyclable lithium and loss of active electrode material

In a non-ideal case, but rather likely in practice, there are combinations of all possible losses, loss of active electrode materials and loss of cyclable lithium, as illustrated in Figure 7.7. These losses include processes such as SEI formation, side reactions, and structural changes in the active electrode materials.

All these losses arise from different ageing mechanisms, strongly dependent on the active materials used for the electrochemical performance, affecting the discharge profiles differently. Modelling the different ageing mechanisms is a useful tool for evaluating the ageing and degradation of full cells for vehicle applications. In

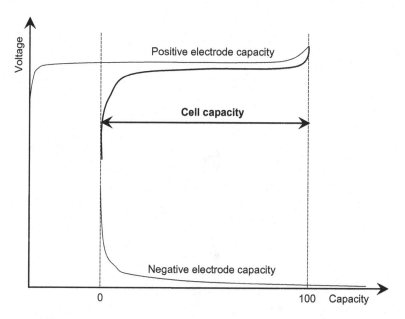

Figure 7.4 Discharge profiles affected by loss of cyclable lithium of a cell and the corresponding electrodes of a tentative Li-ion cell having a larger negative than positive electrode capacity.

combinations with ICA and DVA analysis (Section 7.3), a broader view of ageing will be obtained.

The methods presented can be used to quantify the losses during the life of the cell. In analysing cycle life data by ICA, DVA, and other methods, it is possible to derive useful information regarding the degradation phenomena related to capacity fade or power capability losses. Figure 7.8 illustrates an example of various possible fading behaviours during the cycle life of a cell.

Depending on the chemistry, operational conditions, and age of the cell, the behaviour will vary and therefore no general conclusions can be reached about which degradation phenomenon dominates during the different parts of the cell life. Often, however, the loss of active electrode material will accelerate towards the end of life, and the loss of cyclable lithium is the dominant degradation phenomenon at the beginning of life.

7.1.2   Accelerated degradation

The degradation of a Li-ion cell can be accelerated, unintentionally or not, by non-optimal cycling conditions and temperatures, having a negative impact on capacity fade and power capability. Cycling conditions and temperature correlate strongly; cycling equal cells according to the same procedure but at different temperatures will most likely produce divergent results depending on the temperature differences. The losses in capacity can be temporary or permanent. The former will be recovered once ideal operating conditions are reached and without necessarily having a large impact on the

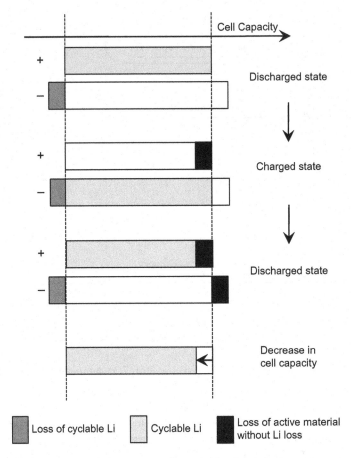

Figure 7.5 Cell capacity decrease during a charge/discharge cycle due to loss of active electrode materials.

durability of the battery. The permanent losses, on the other hand, can be severe and may cause direct damage to the cells, which may even result in abuse conditions, and will always decrease durability.

The power capability and the power fade are dependent on cell resistance and impedance, which in turn are affected by the temperature and the degradation products arising from side reactions. Different active materials will react differently to the accelerating conditions and it is difficult to predict the durability of all types of Li-ion cells by studying only one specific material combination. There are, however, some general parameters that are important to understand accelerating degradation: temperature, cycling conditions, voltage, and current rates.

7.1.2.1   Temperature

Like all battery cells, Li-ion cells have an optimal operating temperature range. Outside this range, at both higher and lower temperatures, cell durability will be reduced

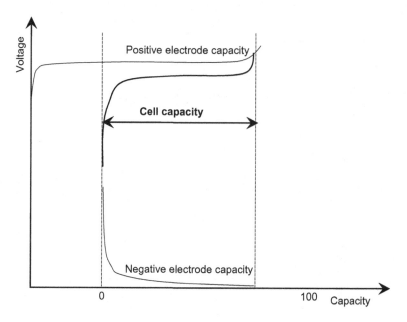

Figure 7.6 Discharge profiles affected by loss of active material in the negative electrode for a tentative Li-ion cell.

dramatically, as schematically illustrated in Figure 7.9. The optimal temperature range varies, however, depending on chemistry and manufacturer, as well as cell format. Even within the optimal temperature range, long durability is not secured since other operating conditions can be dominant for degradation. The operating temperature range can be narrowed to prolong life, which may in turn require more complex and accurate thermal management and control, adding to the overall complexity of the battery and vehicle integration.

As can be seen in Figure 7.9, durability varies with temperature, which is due to different ageing mechanisms taking place. At elevated temperatures, the activation energy of the cell reactions will be lowered, most likely resulting in faster reaction rates. For short periods of time, increased temperature is desirable as the power capability and energy output can increase as a result of faster kinetics. Long-term effects of increased temperature are that degradation also accelerates and side reactions are enhanced, resulting in rapid durability decrease.

As an example, the SEI layer on the negative electrode will start to break and wear down, or dissolve in the electrolyte. As new active electrode material is exposed to the electrolyte, further side reactions will proceed in order to repair the damaged protective SEI layer. This will consume cyclable lithium ions, and cause a decrease in capacity, and subsequently a decrease in durability. An increased temperature will also affect the non-active materials within the cell. The separator may lose some of its functionality, resulting in an increased risk of internal short circuits. Moreover, the composite electrodes will start to deform at elevated temperatures due to softness and decomposition reactions of the binder material. This may affect the electrical contacts between

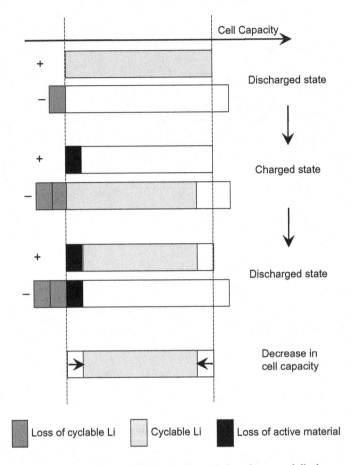

Loss of cyclable Li   Cyclable Li   Loss of active material

Figure 7.7 Cell capacity and lithium balance during charge and discharge of a Li-ion cell including loss of cyclable lithium and loss of active electrode materials.

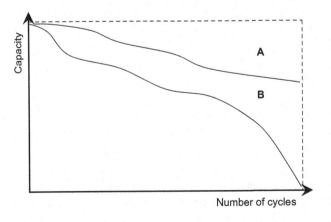

Figure 7.8 Progression of various contributions of capacity fade as a function of the number of cycles. A refers to capacity degradation due to loss of active material, and B to loss of cyclable lithium.

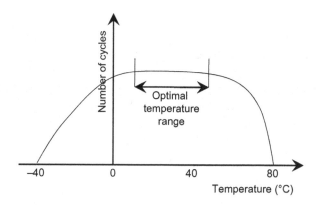

Figure 7.9 Schematic illustration of the optimal operating temperature range for a Li-ion cell with respect to durability.

the particles in the electrode and the contact of the current collectors, reducing the conductivity and adding to the cell resistance. The consequence at battery level will be severe power fade. At too high temperatures (usually above 100 °C), exothermal decomposition reactions could start and may lead to self-sustaining heat evolution, and in the long run a thermal runaway, as described in Section 6.1.2.1.

At lower temperatures (often already below 10 °C), other degradation mechanisms dominate. As the activation energy determines the kinetics, fewer lithium ions will participate in the redox reactions, resulting in loss of cyclable lithium. The lithium ion diffusion is also hampered by lower temperature, leading to reduced power capability of the battery. This capacity loss is though of temporary character and will be overcome once the temperature is increased to the optimal operational range.

Lower temperatures will shorten the life of the battery, as indicated in Figure 7.9, mainly related to charging procedures. The insertion reactions of lithium ions at the negative graphite electrode are slower than the electrical energy transferred, and the electrode will be polarised, resulting in plating of metallic lithium on the electrode surface (Section 3.1.3). This surface layer will block parts of or the entire surface for further insertion reactions, resulting in a drastic capacity loss. In the long run, Li-dendrites may start to grow (Section 3.1.2) further enhanced by a slow diffusion rate of the lithium ions in the electrolyte. The Li-plating is, however, not only caused by low temperatures, as will be seen below.

Another thermal consideration is the thermal cycling. Cells incorporated in a battery for electric vehicles will experience a varied combination of usage and rest periods over ranges of temperatures and SOC, as well as seasonal and geographical variations. These variations may accelerate some of the degradation mechanisms and the capacity loss and resistance increase may become worse when a Li-ion cell is stored or cycled at different temperature conditions. When stored, a battery thus clearly benefits from having a uniform and controlled temperature. Since most degradation is strongly related to the usage and storage history, and is path-dependent, it is useful to know the storage conditions for any durability prediction.

7.1.2.2   Voltage

Different types of electric vehicles utilise different SOC ranges (Section 5.1), i.e. within specific voltage limits for the given battery technology. How to select the optimal SOC range can be crucial for durability. High and low SOC levels will have an impact on durability, and preferably the battery should be operated within a specific SOC range, often set by the cell manufacturer. It is, however, always possible for the battery to operate outside this range, due to usage demands and to unwanted extremes like overcharge and overdischarge states. The latter two will accelerate the cell degradation and therefore be used to illustrate how extreme voltages can affect cell life.

At overcharge conditions, i.e. at voltages above the upper cut-off voltage, electrical energy is forced into the battery but no further lithium ions can be extracted from the positive electrode, which will result in a rapid increase of the internal cell resistance. The forced energy will be converted into heat with consequences as described above. The overcharge will also influence the materials of the electrolyte and the electrodes. At overcharge, or high SOC levels, a substantial part of the cyclable lithium ions is found in the negative electrode. Depending on the active material used for the positive electrode, the depletion of lithium from the positive electrode can cause deformation reactions of the active electrode material due to the thermodynamic instability of the material. This can result in severe long-term damage to the electrode, and further lithium insertion during the subsequent discharge of the cell may be impossible. These structural changes may also cause abuse situations, depending on the chemistry of the active material. As the voltage increases and surpasses the upper limit of the electrochemical stability window (Section 1.5.6), the electrolyte starts to decompose and form insoluble products, blocking the electrodes, or gases, which at worst may cause cell rupture and safety hazards.

During overdischarge, i.e. when the voltage has decreased below the lower cut-off voltage, other degradation mechanisms take place, mainly related to the negative electrode, and severe capacity fade or abuse situations may arise. At these low voltages, the polarisation effects will force the SEI layer on the negative electrode to start decomposing and dissolving. This in turn will initiate reactions to restore the SEI layer and consequently a loss of capacity will be observed. Furthermore, the copper foil current collector of the negative electrode will corrode, and this may be the most severe degradation mechanism. As a consequence, the contact resistance between the current collector and the composite electrode will increase due to the contact losses, resulting in power fade. Furthermore, the copper ions will diffuse towards the positive electrode and could block the active materials for further insertion/extraction reactions, and in the worst case the subsequent charging may cause the formation of a conduction path of dendrites, via the separator, causing an internal short circuit.

7.1.2.3   Current rates

The most important of the operational parameters is current, and especially the current rates. The cycling of the battery will proceed at different current rates and in different directions, i.e. charge or discharge conditions. Often the C-rate is used, a function of the current related to the rated capacity of the cell. As for Li-ion cells applicable to electric vehicles, a high C-rate is often in the range of 5–10C, and a low rate less than 1C, but they differ depending on cell chemistry and cell design constraints. Moreover, the C-rate is also related to the power capability of the cell, i.e. if the cell is optimised for energy or power. As shown in the cell selection example in Section 5.2.2.2, the maximum C-rate of a power-optimised cell is significantly higher than of an energy-optimised cell. Therefore, in the following sections, high and low C-rates cannot be specified by numbers, but the phenomena caused by high currents are more or less general both for energy and power-optimised cells.

High C-rates will deplete the lithium ions available for insertion reactions near the electrode surface since the diffusion of the lithium ions in the electrolyte is too slow, causing a concentration gradient to develop near the electrode surface (Section 1.5.3). Even if lithium ions are available in the bulk of the electrolyte, they are not accessible to the redox reactions. This gives rise to increased cell impedance, and if this proceeds for a long time, the cell would be damaged due to local overpotentials. After relaxation, the concentration gradient will disappear as new lithium ions diffuse to the electrode surface and the insertion reactions continue. The capacity loss is therefore temporary. At low C-rates, on the other hand, the diffusion of the lithium ions will always be fast enough to minimise the concentration gradient between the electrode surface and the bulk of the electrolyte.

At high discharge rates, the slow diffusion rates of lithium ions in the electrolyte and in the active electrode materials will cause concentration polarisation to increase, as well as ohmic and activation polarisation (Section 1.6.1), resulting in a voltage drop, a voltage drop related to the C-rate of the discharge procedure. As the voltage drops, the lower cut-off voltage will be reached faster, and consequently the capacity and power fade will be more obvious.

For high charge rates, the opposite is true: the upper cut-off voltage will be reached earlier and a lower charge capacity will be obtained. High C-rates may also cause strain in the composite electrodes, resulting in cracking both of the active electrode materials and protective SEI layer. These mechanisms may further result in increased cell resistance, severe loss of accessible active material, and a decreased amount of cyclable lithium ions. Both charge and discharge will thus cause polarisations and losses, the energy efficiency will be lower at higher C-rates, and these energy losses will mainly be transformed into heat. This even further illustrates the interactions between the different operational parameters, here C-rates and temperature.

Usually short pulses, less than seconds, of high C-rates will only cause temporary capacity losses as the cell is able to recover and the lithium ion diffusion to the electrode surface can catch up to prevent high concentration gradients and local overpotentials. On the other hand, using high C-rates over long periods of time will damage the cell and affect durability negatively as the Joule heating, or ohmic heating (Section 1.4.3), increases the temperature of the cell. Consequently, a continuously high, charge or discharge, current will cause a temperature rise and may lead to the degradation mechanisms described above. If the thermal capacity of the cell is large, the generated heat may be negligible, but otherwise the thermal system of the battery must be able to remove the excess heat.

7.2   Degradation of Li-ion cells

The degradation mechanisms and the factors affecting degradation will now be put in the context of materials inside Li-ion cells. It is more or less impossible to find a single battery cell serving all-electric vehicle applications, and in the same way it is difficult to find one degradation model describing the ageing of all types of Li-ion cells. In electric vehicles, the primary ageing factors are temperature, voltage, and current, all affecting the materials and initiating degradation reactions. The ageing is a complex field of battery understanding and design, and the ageing of Li-ion cells originates from various degradation mechanisms of the materials used, caused by the cycling and storage conditions.

As discussed previously, most degradation processes are connected in several ways, making the process to assign the observed capacity or power fade to a specific usage condition difficult. In the case of extreme conditions, however, it might be possible to separate the observations and attribute it to a specific usage condition.

Knowledge about, and understanding of how, the Li-ion cell materials are affected by temperature, voltage, and current will be the input for all battery management systems in order to use the battery in the most optimal way: to secure performance, safety, and life. Knowledge about the different materials in the Li-ion cell is essential, as the various materials will degrade differently. As a consequence, it is the ability to avoid operational conditions triggering the degradation reactions that makes it possible to achieve a long battery life. This ability to manage the battery is of high importance to the competitiveness of the vehicle manufacturer, and would result in EVs having long and stable all-electric driving ranges, and HEVs having profitable fuel economy during the life of the vehicle.

A number of different processes, related to electrochemical, mechanical, and cell design, often in combination with each other, affect the life and durability of the cells, and consequently the performance of the electric vehicle. Most of the degradation mechanisms of a Li-ion cell originate at the electrodes and at the interfaces between the electrodes and the electrolyte. Both the active and the non-active materials are subjected to ageing at different degradation rates, in the end rendering the cell useless. The active material may expand during usage inducing mechanical stress resulting in loss of electrical contacts within the electrodes. Mechanical distortion of the cell components may cause short circuits or open circuits, which may result in severe abuse conditions. Many degradation processes are due to irreversible chemical reactions, causing permanent reduction in the active materials and lowering the cell capacity. The temperature, both the ambient and internal caused by the chemical reactions, may enhance the rate of the side reactions, which may generate pressure build-up within the cell, and in the case of cell rupture, toxic and flammable chemicals may be released.

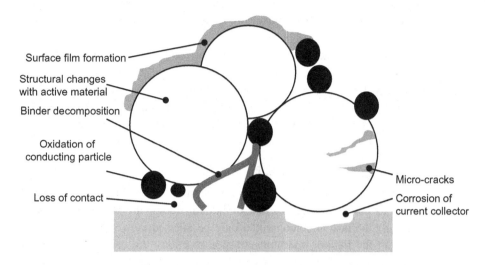

Surface film formation

Structural changes
with active material

Binder decomposition

Oxidation of
conducting particle

Loss of contact

Micro-cracks

Corrosion of
current collector

Figure 7.10 Main degradation processes occurring at the electrodes of Li-ion cells.

The degradation mechanisms have several similarities independent of the materials selected, as illustrated in Figure 7.10.

7.2.1 General degradation categories

The processes of Figure 7.10 can be grouped into the following general categories:

- mechanical and electrical changes
- changes in bulk materials
- surface film formation
- side reactions

7.2.1.1 Mechanical and electrical changes

The cell must have mechanical stability throughout its life in order to achieve reliable performance. During operation, the electrodes will experience mechanical stress caused by insertion and extraction of lithium ions, affecting the electrical conductivity between the particles of the composite electrodes. Furthermore, as a consequence of the cycling conditions, the particles of the active material in the electrodes may form agglomerates, resulting in reshaping of the structures and porosity of the electrodes. The changes in particle size and morphology could give rise to changes in the electrical contacts within the electrodes.

Depending on the electrolyte utilised, solvent molecules may be co-inserted into the graphite-based negative electrode, causing structural degradation of the graphite host, and consequently loss of active material and loss of capacity. Micro-cracks may also appear during cell operation. The cracks will result in an increased surface area of the electrodes, and an increased interface towards the electrolyte, and possible decomposition reactions between the two. The cracks may propagate and the complete particle may burst and further contact resistance may occur due to loss in electrical contact between the particles.

Loss of contact between the particles, both the active material and the carbon-conducting particles, as well as the contact with the current collector, will result in increased internal resistance and loss of the power capability of the cell. The increased resistance is also caused by the exfoliation of the composite electrode from the current collectors. Severe loss of contact between the electrode material and the current collector could result in hot spots and damage due to locally increased temperature.

7.2.1.2   Changes in bulk materials

The properties of the bulk material in the positive and the negative electrodes change due to the insertion/extraction of lithium during the charging–discharging process. Ideally the host structure would only be affected by a volume expansion/contraction during the lithium insertion/extraction reactions. As described above, though, these volume changes result in mechanical stress and may cause cracking of the particles, which, in turn, may cause contact losses and increased resistance. The host structure may also change structurally and can result in new materials possessing other electro-chemical properties. These changes are mainly related to the positive electrode and they are caused by forced lithium insertion or extraction beyond the limits of the material, i.e. overcharge or overdischarge conditions. The new structures formed can be active or inactive in terms of lithium insertion/extraction or accompanied by decomposition reactions, e.g. gas evolution or dissolution of material into the electrolyte. The structural disorder of the host structure can also be a result of local temperature changes.

7.2.1.3   Surface film formation

Many degradation mechanisms arise at the interface between the electrodes and the electrolyte. As described in Section 1.5.6, the electrolyte must be stable over a wide voltage range. Depending on the combinations of electrode–electrolyte, a film is often formed on the electrode in order to protect the surface from further degradation. The film is made of decomposition products mainly from the electrolyte and this is important for the graphite negative electrode (Section 3.1.3), the most widely used negative electrode in commercial Li-ion cells, since the graphite is not electrochemically stable in the most common electrolytes. A surface film may also be formed on the positive electrodes by equivalent mechanisms and consequently a loss of cyclable lithium and increased power fade will be observed.

A temperature increase may partly dissolve the surface films and the electrode is then again exposed to the electrolyte, forming a new film consuming even more cyclable lithium. Moreover, micro-cracks appearing during cycling will also cause cracks in the surface films and consequently a new film will be formed. The loss of cyclable lithium will have long-term effects on the cells resulting in capacity loss.

7.2.1.4   Side reactions

All degradation reactions are parasitic in some sense. Degradation reactions not related to the active materials are side reactions. One part of the electrode mainly generating mechanical changes and film formation is the binder material holding the electrode together. The material used may degrade and film-forming compounds, possibly in gas form (mainly CO_2), may be produced. A loss of binder material will, however, mostly affect the electric contact between the particles and the mechanical stability of the electrode, thereby reducing the power capability of the cell due to increased resistance. Lithium ions may also be directly involved in the side reactions with electrode and electrolyte species to form soluble and insoluble products, causing irreversible loss of cyclable lithium. The soluble products are mainly involved in the self-discharge reactions, and the insoluble in surface film formation.

Material impurities and process contaminations originating from the production and assembly processes are the sources of many side reactions. In Li-ion cells, water is one of the more common contaminants reacting with the salt in the electrolyte to form HF. Most likely the HF formed will etch the electrodes and destroy the protecting SEI layer, enabling protons of HF to penetrate further through the electrodes in order to react with the current collectors. Side reactions can also arise from contamination at one of the electrodes and be dissolved upon cycling and transported through the electrolyte to form seeds on the opposite electrode for further reactions. These contaminants may also block the pores of the separator material, inhibiting the ion transport and thereby limiting the rate capability of the cell. Moreover, contaminants may cause dendrites to grow, which could result in internal short circuits and hazardous events. Another type of side reaction is the corrosion of the current collectors, which may take place if exposed to potentials close to or exceeding the electrochemical stability potential.

Another type of side reaction, not of the same character as those above, is lithium plating (Section 3.1.3): metallic lithium being plated on the surface of the negative electrode, resulting in loss of cyclable lithium, and thereby a loss of capacity. Lithium plating by itself is a reversible reaction, but the plated lithium may remain at the electrode surface due to unfavourable operating conditions, and thus increase cell resistance, thereby limiting the power capability of the cell. Moreover, a surface film will be formed on the plated lithium as metallic lithium is unstable in the organic electrolyte.

An additional stress factor of the cell is gas evolution as a consequence of side reactions and this may have severe abuse consequences as the welds of the cell may rupture allowing gases to be emitted. Therefore, the cells are often equipped with safety vents or a CID (Section 4.4.4). The gases may be toxic and/or flammable and could cause severe hazard situations, like fires or explosions. For pouch cells, the consequences may be less severe due to the weaker welding junctions.

The most common degradation mechanisms occurring in Li-ion cells are summarised in Table 7.1. The causes enhancing the degradation mechanisms and the corresponding effect on the performance are given. Some of the mechanisms are only encountered at the negative electrode, e.g. Li-plating and co-insertion of solvent, but most of the mechanisms are valid for both electrodes.

Table 7.1 Causes and effects for the most common degradation mechanisms associated with Li-ion electrodes.

Cause	Effect	Result in	Increased by
Changes in particles of electrodes	Increase of impedance and resistance	Power fade	High C-rates
Contact losses between active material and current collector	Increase of impedance and resistance	Power fade	High C-rates
Cracking of electrodes	Loss of active material	Capacity fade	Overcharge High C-rates
Structural changes in active materials	Loss of active material	Capacity fade	Overcharge
Electrolyte decomposition and SEI formation	Loss of cyclable lithium and increased impedance	Capacity and power fade	
Decomposition of binder	Loss of lithium and mechanical stability	Capacity and power fade	High temperature
Current collector corrosion	Increase of impedance and resistance	Power fade	Overdischarge
Li-plating	Loss of lithium	Capacity fade	High C-rates Low temperatures
Gas evolution	Loss of active material	Capacity fade	Overcharge

7.2.2   Degradation of active materials

So far, degradation mechanisms have been discussed in general terms. The various cell materials, active as well as non-active, undergo different degradation and ageing mechanisms: e.g. the binder and the electrolyte decompose, the current collectors corrode, the separator changes structure and composition, the active positive electrode material changes structurally and metal dissolves, and the active negative electrode material forms passivation films.

Many of these mechanisms are dependent on the microstructure, crystallinity, and morphology of the active electrode materials, which are influencing the cell perform-ance directly. In the following sections, specific degradation aspects of the composite electrodes, the negative and the positive, and the electrolytes used in the Li-ion cells will be described from a materials perspective. The most common and promising materials will be discussed with reference to the degradation phenomena occurring for that specific material during operation and storage of the cells.

7.2.2.1   Degradation of negative electrodes

The degradation mechanisms described above are certainly in general valid for the negative composite electrode. In order to illustrate in more detail the degradation phenomena occurring at the negative electrode, a graphite-based electrode will be used.

The vast majority of the degradation of the negative composite electrode takes place at the interface between the electrode and the electrolyte. The electrode is outside the electrochemical stability window during operation and, hence, reductive electrolyte decomposition will occur. This is accompanied by an irreversible loss of cyclable lithium and a surface layer will be formed on the electrode surface during the first charge cycle of the cell. The graphite electrodes will be polarised at these low potentials and will be covered by the SEI layer. As described in Section 3.1.1, the SEI layer is necessary for overall cell performance, while it gives rise to increased impedance of the cell.

The SEI layer prevents continued electrolyte degradation by blocking the electron transport and at the same time allowing lithium ion diffusion. The formation of the SEI layer on the graphite electrode is thus both a protective measure and a predominant source of cyclable lithium loss, resulting in increased charge-transfer resistance, increased impedance, and limited access to the anode surface. The result is an increased irreversible capacity loss. Depending on the particle size and morphology of the electrode surface, different amounts of SEI will form, and thereby the loss of cyclable lithium ions will vary. The SEI can penetrate the pores of the electrode to inhibit the contacts between the particles and the conduction paths, resulting in a reduced accessible active surface of the electrode.

The SEI formation and electrolyte decomposition are ongoing processes throughout the entire life of the cells. The growth rate is highest at the beginning of life, and then it declines rapidly. It is possible to inhibit the growth rate, but it is impossible to fully eliminate it. The SEI is a dynamic and inconstant part of the Li-ion cells; the thickness, composition, and morphology change with time and use, increasing the impedance of the cell and thereby reducing the power capability. The content of the SEI layer is heterogeneous and varies depending on the material properties of the electrodes, as well as the electrolyte. The most common species are Li_2CO_3, LiF, Li_2O, $LiOH$, polycarbonates, and lithium-containing organic species. Some additives, as described in Section 3.3.1.3, are solely acting to reduce the growth rate of the SEI. Moreover, the different kinds of graphite material used in the electrode preparation will influence the growth and stability of the SEI.

The interaction between the negative and the positive electrodes has to be considered in the degradation of the negative electrode. Depending on the positive electrode used, dissolution of the active electrode material may occur. Dissolved transition metal ions (e.g. Mn^{2+}, Fe^{2+}) can be transferred through the electrolyte and incorporated in the SEI layer, resulting in loss of active material at both the negative and the positive electrodes.

Temperature has an effect on the SEI layer thickness, composition and morphology. At increased temperatures, the SEI starts to break down or dissolve. The SEI layer may also dissolve during the cycling of the cell and also the C-rate will affect the SEI dissolution and precipitation. The SEI dissolution in the electrolyte may form insoluble particles, which may, for example, block the pores in the separator, and thereby increase cell resistance. During storage, the SEI layer may grow, causing higher self-discharge rates and consequently irreversible capacity losses. Moreover, storage at elevated temperatures may lead to crystal growth of LiF, which could react with traces of water to form the highly reactive HF.

At low temperatures, large polarisation occurs during the insertion of lithium into graphite, forcing a deposition of lithium on the surface before insertion can take place. This deposition, lithium plating (Section 3.1.3), on the electrode surface causes an overpotential during the discharge process, and can be observed in the discharge profile (Figure 3.10). However, the plated lithium may remain at the electrode surface due to unfavourable operating conditions, and will thus increase cell resistance. In addition, if the current is non-uniformly distributed over the electrode surface it may also cause dendrites to form. Lithium plating and dendrite growth will change the interfacial conditions of the SEI layer and the surface of the negative electrode.

Also the bulk of the active graphite material may undergo volume and structural changes during cell ageing, mainly resulting in impedance rises and capacity fading. The stacking order of the graphene layers within graphite can be shifted during the lithium insertion and extraction reactions (Section 3.1.3). During charge and discharge, the graphite structure will expand and contract due to lithium insertion and extraction. This could cause mechanical stress on the carbon–carbon bond, which in the worst case results in cracking. The SEI layer must then be able to accommodate the same volume changes as the graphite, otherwise cracks may be formed also in the SEI, consequently exposing a fresh graphite surface triggering the formation of a new SEI layer which consumes more cyclable lithium and reduces cell capacity. Degradation related to changes in the bulk material is most often negligible compared to that related to the SEI layer.

The electrolyte may also be part of the degradation of the negative electrode. Depending on the electrolyte composition, solvent molecules can be co-inserted into the graphite structure resulting in exfoliations of the graphene sheets and cracking of the particles. The co-inserted solvent molecules will be reduced within the graphite structure, and during this decomposition, gaseous species, mainly CO_2, may evolve, which, in turn, may cause further cracking and exfoliation of the active graphite material.

As for processes involving the inactive materials, corrosion of the current collector and decomposition of the binder material are sources of impedance rises, and consequently decreased power capability of the cell due to the loss of electrical contacts within the composite electrode and with the current collector. The increased contact resistance may also lead to locally increased temperatures, enabling further degradation reactions to occur. The corrosion product from the current collector, the copper foil, will also cause other degradation mechanisms within the cell.

At overdischarged states, copper can dissolve and copper ions may diffuse throughout the composite electrode and precipitate on the surface of the electrode. Moreover, the Cu current collector is not stable if traces of HF are present in the electrolyte and acid-base reactions will take place between the protective oxide layer and the HF impurities. Different layers for protection and for improvement of the electrode adhesion to the current collectors can be used, but as for all non-active materials added to the cell they should be compatible and stable in relation to the electrochemical environment in the cell.

The dominant degradation mechanisms occurring at the negative graphite electrode are summarised and illustrated in Figure 7.11.

The SEI layer traps cyclable lithium and decomposed electrolyte species, and the SEI thickness is affected by temperature and C-rates. Morphological changes and structural reordering of the active graphite particles will influence SEI layer stability and the

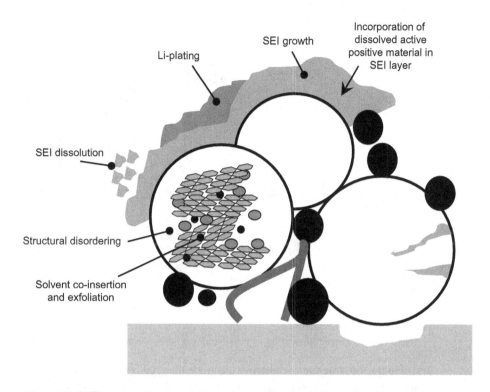

Li-plating

SEI growth

Incorporation of
dissolved active
positive material in
SEI layer

SEI dissolution

Structural disordering

Solvent co-insertion
and exfoliation

Figure 7.11 Different possible degradations of a negative electrode with graphite as active negative electrode material.

insertion processes. The co-insertion, and the subsequent decomposition, of solvent molecules will lead to exfoliation of the graphene layers of the graphite. Loss of electrical contact between the graphite particles and the current collector is a severe failure of the electrode performance. Finally, the lithium plating is a source of lithium loss, and a safety related issue. It should be kept in mind that these degradation mechanisms are general and there might be pronounced differences in each particular cell chemistry used, especially due to the composition of the electrolyte. These degradation mechanisms will give rise to capacity fade and impedance rise of the cell. Most of them are sensitive to both high and low temperatures, being constrained by the thermal system design of the battery.

7.2.2.2   Degradation of positive electrodes

Many of the degradation mechanisms occurring at the positive electrode are similar to those of the negative electrode. In general, the cell capacity is mainly reduced due to (i) structural changes during lithium insertion/extraction, (ii) dissolution of active material, and (iii) surface reactions. These degradation mechanisms of the positive electrode are highly sensitive to the SOC of the cell, i.e. the lithium content in the active material. Possible degradation mechanisms of the positive electrode are summarised and illustrated in Figure 7.12.

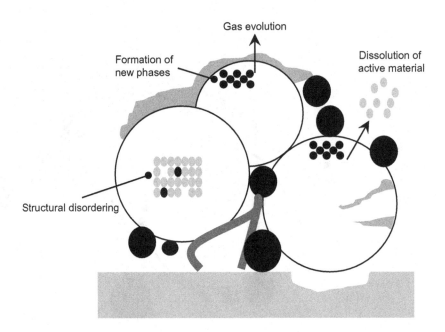

Figure 7.12 Schematic illustration of possible degradation mechanisms in the positive electrode.

Firstly, the structure of the active electrode material is the main source of degradation, mostly related to structural disordering or phase transitions causing mechanical stress in the material. These degradation mechanisms result in voltage and capacity fade and hence loss in cell performance. The electrochemical performance of the disordered structures or the new phases obtained by phase transition can be limited or absent. In such cases, the lithium kinetics within the solid materials are reduced. Hence, cell impedance is increased, and consequently power capability is decreased.

Secondly, metal ion dissolution from the metal oxides, often used as the active material, may also result in new phases. This results in similar issues as for structural disordering or phase transitions. Furthermore, the dissolved metal ions can diffuse through the electrolyte to be incorporated in the SEI layer of the negative electrode, which will result in capacity fading.

The third degradation mechanism is related to the surface of the electrode. As in the case of the negative electrode, the electrolyte may decompose at the electrode surface with subsequent film formation, resulting in an increase of the cell impedance. Furthermore, the electrolyte decomposition can also involve gas-evolving reactions and often depend on the potential of the positive electrodes; at high electrode potentials, the upper limit of the electrochemical stability window can be reached. The gases formed, e.g. CO_2, CO, and gaseous hydrocarbons, will increase the internal cell pressure, and could cause venting of the cell.

Finally, just like the copper current collector of the negative electrode, the aluminium current collector of the positive electrode is affected by a corrosive environment. The Al current collector is constantly held at potentials where it is thermodynamically unstable. Therefore, a thin protective film of various Al_x-O_y-F_z compounds are applied to the surface in order to prevent corrosive pitting of the current collector at potentials above *ca.* 3.6 V vs. Li/Li$^+$. Depending on the salt used, the protective surface film on the Al current collector becomes unstable and corrosion processes occur. Aluminium ions can then be dissolved and transported to the negative electrode via the electrolyte, where they may impair the performance of the negative electrode by blocking the lithium diffusion pathways or interfering with the SEI layer.

To further illustrate the degradation mechanisms of the active materials of the positive electrode, specific degradation reactions will be exemplified by some active materials affected differently by ageing: the layered LiCoO$_2$-based materials NCA and NMC, the cubic spinel LiMn$_2$O$_4$, and the olivine LiFePO$_4$. The structures and electrochemical properties of these materials are described in Section 3.2. The materials will be described with respect to degradation mechanisms like overcharge and overdischarge, thermal instability, and the sensitiveness towards water contaminations and impurities.

Layered LiMO$_2$ materials

As described in Section 3.2.1, there are several types of layered positive electrode materials based on LiCo/NiO$_2$ where substitution with aluminium or manganese is used to stabilise the base structure and form the materials NCA and NMC. Both these possess electrochemical performance characteristics of interest to electric vehicles. The structure basically enables two-dimensional lithium diffusion pathways, but depending on the specific composition of the NCA and NMC materials, disordering of the cations may take place hampering the lithium diffusion paths, and consequently reducing the capacity and the power capability of the cell.

The materials are based on structures being sensitive to both temperature and high potentials, i.e. the amount of lithium within the structure. At potentials above 4.2 V vs. Li/Li$^+$, i.e. small amount of lithium within the structure, a surface layer of decomposed electrolyte products will be found on the positive electrode. This is similar to the SEI layer on carbon-based negative electrodes, due to the potential being above the LUMO level and thereby above the upper limit of the electrochemical stability window. The products formed on the surface are of organic character, e.g. dicarbonyl and ester compounds, and they are not found at lower potentials. Some of the reactions are accompanied by evolution of gases. These layers will increase the polarisation resistance of the cell, and a rise in impedance will be observed. Moreover, the oxides themselves can act as sources of oxygen for the electrolyte decomposition reactions by forming additional surface layers of oxygen deficient oxides and subsequent reduction of the transition metals.

The reactions are associated with structural changes from a 2D to a 3D structure of the surface of the active material particles and at the grain boundaries. These new structures often exhibit reduced electrochemical performances leading to an impedance rise due to reduced lithium kinetics. The reactivity of the surface particles can be high in the presence of electrolyte. The oxidation state of nickel at the particle surface will be reduced, forming inactive Ni-O phases. Since they are not involved in the insertion/extraction reactions, a loss in capacity will be obtained. Moreover, the Ni-O formation occurring during the lithium extraction process will release oxygen, causing internal pressure build-up with continuous cycling. Electrolyte additives may be added in order to limit surface reactivity of the active electrode material.

The most severe degradation is due to thermal stability. In the delithiated state, the positive electrode materials are highly exothermal at temperatures close to the onset temperature of structural changes. The Me-O bonds are of covalent character and relatively weak, and decomposition occurs at overpotentials or at temperatures above *ca.* 170 °C. Due to the active material incorporated in Li-ion cells, the thermal increase can, however, be fast, e.g. creating hot-spots or internal short circuits, and could trigger thermal runaway. Therefore, careful control of the SOC, as well as the temperature, is required for NCA and NMC-based Li-ion cells.

Cubic spinel $LiMn_2O_4$

The cubic manganese oxide spinel structure $LiMn_2O_4$ (Section 3.2.2) is accompanied by degradation mechanisms other than NCA and NMC, and is particularly sensitive to the SOC level and to impurities in the electrolyte. The main degradation mechanism is the dissolution of active material and this may occur at both high and low SOC. The consequence is loss of active material and a corresponding loss of capacity. The dissolution reactions are related to the Mn^{III} in the structure and could be crucial especially at elevated temperatures, putting further constraints on the thermal management of the battery.

When approaching a fully discharged state, the oxidation state of manganese comes close to III, and the structure starts to disproportionate; the trivalent manganese disproportionates into tetravalent and divalent manganese according to:

$$2MnI^{III} \rightarrow Mn^{IV} + Mn^{2+}$$

The Mn^{2+} ions are subsequently dissolved in the electrolyte, whereas Mn^{IV} remains in the solid spinel structure. This degradation reaction will not only cause a loss of active material of the positive electrode, but will also affect the negative electrode. Solvated Mn^{2+} may be transported through the electrolyte and become incorporated in the SEI layer, acting as impurity phases enhancing the decomposition reactions of the SEI layer. As these reactions are related to the oxidation state of the manganese atoms in the spinel structure, discharging of Li-ion cells based on the spinel material towards low SOC levels should be avoided to prevent disproportional reactions.

The dissolution of manganese can, however, occur at higher SOC levels as well, caused by chemical reactions involving mainly HF. The source of HF is primarily hydrolysis of the $LiPF_6$ salt used in the electrolyte induced by, for example, water impurities. The dissolution reaction involving HF is:

$$Li_{1-x}Mn^{III}Mn^{IV}O_4 + HF \rightarrow \lambda\text{-}Mn^{IV}O_2 + Mn^{2+} + (1-x)LiF + H_2O$$

Another degradation mechanism caused by overdischarging the cell is the formation of $Li_{1+x}Mn_2O_4$, accompanied by a structural transition. This is due to the dominance of Mn^{III} in the structure resulting in a Jahn–Teller distortion where the octahedral coordinated Mn^{III} changes from cubic to tetragonal symmetry. The large anisotropic volume change associated with the formation of $Li_2Mn_2O_4$ introduces stress in the material and loss of electrical contacts within the electrode, resulting in losses in both the capacity and the power capability of the cell. This causes such a rapid loss of capacity that, in practice, the cycle life of $LiMn_2O_4$ based cells overdischarged only a few times is greatly compromised. Therefore, the battery management system must be able to control the discharge of the cell in order to prevent any overdischarging.

Olivine LiFePO$_4$

The final example of degradation mechanisms for the positive electrodes is given for the olivine $LiFePO_4$ material. Cells based on high-quality produced $LiFePO_4$ will, however, degrade almost exclusively due to degradation of the negative graphite electrode and the loss of cyclable lithium is the dominant factor responsible for the capacity loss of the cell. The stability of the $LiFePO_4$ material is dependent on the synthesis process, as well as the assembly of the cells. The synthesis route and cell processing conditions must be under strict control to limit the contaminants, mainly water and Fe-based particles.

The diffusion pathways within the structure are one-dimensional, and the material has limited power capability due to low electronic conductivity of the solid structure. Any structural distortions originating from material synthesis can further reduce the power capability by blocking of the pathways, which may trap lithium in the structure causing loss of cyclable lithium and consequently a reduction of the cell capacity as well.

Even if the material is stable, water impurities can affect its long-term performance due to dissolution of Fe^{2+}, especially in the presence of protic contaminants (similar to the dissolution reactions of $LiMn_2O_4$ described above), and Fe-based impurities present in the electrode may further contribute to the dissolution reactions. Migration of the dissolved ions to the negative electrode through the electrolyte is likely. At the negative electrode, the ions can be reduced to metallic iron, causing an impedance rise of the negative electrode and subsequently capacity fading and loss of the power capability of the cell. Moreover, the metallic iron can lead to dendrite growth, and a possible internal short circuit in the worst case. The iron dissolution can be limited by using water-free cell materials and preparation methods.

The thermal stability of the $LiFePO_4$ material, up to about 1000 °C, is high due to the strongly bound oxygen atoms in the phosphate group, and the resulting decomposition reactions are endothermic. If, however, the delithiated $FePO_4$ phase is exposed to high temperatures (*ca.* 350° C), an irreversible phase-transition takes place. This occurs without any release of O_2 gas due to the strongly bound oxygen atoms within the phosphate anion.

7.2.3   Degradation of electrolytes

Electrolyte degradation is mainly related to the decomposition of the solvent and salts, but also coupled to the separator used for Li-ion cells based on liquid electrolytes. Most of the degradation reactions take place at the surface of the electrodes, especially at the negative. These reactions will consume electrolyte mainly for the formation and growth of the SEI layer. As the amount of electrolyte is reduced, it will impair the diffusion and kinetics of the lithium ions resulting in increased resistance and power fading. Parts of the separator can be impoverished causing increased resistance and temperatures locally, and thermal runaway or internal short circuits may occur. One way to understand the degradation rates and propagation is to measure the amount of free electrolyte of an aged cell compared to a fresh one; impossible, however, in vehicle applications.

The salt used, usually $LiPF_6$, is sensitive to water contaminants, and depending on the amount of water the salt can decompose to form HF, for example, according to:

$$LiPF_6 + H_2O \rightarrow LiF + 2HF + POF_3$$

$$LiPF_6 + 4H_2O \rightarrow LiF + 5HF + H_3PO_4$$

Besides the purity of the raw materials used, the quality and process control of the electrolyte synthesis and cell production strongly influence the durability of the battery. Electrolyte additives can, however, be used in order to neutralise HF and absorb the protons formed.

If the electrolyte is subjected to conditions leading to decomposition, the organic solvents are very likely sources of various gases evolving (e.g. CO, CO_2, $C_xH_yF_z$ compounds) and also of exothermal, i.e. heat-generating, reactions. Such events can in turn trigger further decomposition of, for example, the SEI layer, as well as the composite electrodes. The electrolyte being outside its stability range is often the major cause for Li-ion cell degradation and abuse conditions.

7.2.3.1   Separator degradation

As indicated above, electrolyte degradation and failure can be coupled with and due to the separator. The degradation of the separator will proceed during cell usage, as well as storage, and the degradation can be classified as mechanical or thermal, either case causing increased cell impedance and thereby power losses. The increased cell impedance is due to the decreased lithium ion conductivity through the separator as pores are blocked.

Throughout cell life, the mechanical strength and the structural stability of the separator are necessities, and its structure should remain and not crumple even when soaked in electrolyte to keep the interface between the electrodes and the separator intact. Pinholes can be present in the separators as defects arise from the manufacturing and cell assembly processes and these pinholes can affect the self-discharge rate. Moreover, a high-impedance cell connected in series could be subjected to a higher voltage environment from the other cells, which may lead to a breakdown of the separator.

Temperature is the primary cause of the degradation, especially elevated temperatures. The increased temperature will cause melting processes or structural changes to occur. As the temperature rises, the separator reaches its softening temperature at which the separator tends to shrink, even at low porosity. The melting process may, however, be used as a safety feature of the cell; the charging process can be terminated internally, preventing abuse situations like thermal runaway. Thermal gradients within the battery will give rise to variations of cell impedance across the battery as a result of the softening and structural changes in the separator. Once again, the design and control of the thermal management of the battery are of utmost importance in the extension of battery life.

7.3   Degradation analysis methods

To test, evaluate, and map the actual battery life calls for a large test matrix of all possible operating conditions. This will, however, require a large number of test objects and be as time consuming as the actual use of the vehicle, which is unacceptable in the design process. Therefore, accelerated test procedures are used.

Since it is not possible to measure the SOH of a battery (only estimate and predict) related to the conditions at BOL, different analysis methods are used to obtain adequate parameters for mathematical models used to predict battery durability. The models are based on experimental data from different calendar and cycle life tests, or on fundamental electrochemical models, including mass-transport and reaction kinetic constraints. Data from both battery and cell measurements are often preferred to obtain reliable life predictions. Apparently, it is difficult, or more or less impossible, to find a single method to validate all battery technologies, cell formats, and operational conditions. But first a bit about test conditions – crucial for analysis methods.

Cell and battery cycling
Before any analysis can be made, the tested cells have to be cycled. Although the ageing mechanisms may be similar, these initial procedures, i.e. the test conditions, will highly influence the resulting life estimations due to the non-linearity of the degradation processes. Thus, small changes in test conditions may cause one or several degradation processes to dominate in a way not reflecting the actual vehicle battery usage. Generally, there are two ways of testing the life of a battery: either according to standardised procedures or by representing procedures according to the specific type of vehicle.

The standardised tests often employ full charge/discharge cycles (e.g. galvanostatic cycling, Section 1.7.1) and are used for comparing different cells or cell designs rather than determining cycle life. Often these standardised tests are performed at elevated temperatures in order to accelerate ageing. Due to the increased rate of the chemical reactions at elevated temperatures, of both redox and side reactions, this procedure is not recommended unless the products of the side reactions are of interest.

Using charge/discharge profiles reflecting the actual usage conditions of a specific vehicle will often result in a better understanding of the battery degradation and corresponding life. To select representative usage conditions is not a straightforward process and there are many parameters to consider, e.g. SOC level and range, temperature, C-rates, and the energy throughput. One way is to use data from a vehicle during real driving operation. This will, however, pose a challenge in terms of comparing different cycles. Due to the complexity of battery operation, accurate diagnosis and prognosis are vital.

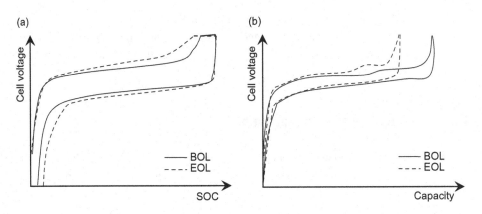

Figure 7.13 Discharge profiles vs. SOC (a) and capacity (b).

Cells from batteries already used in vehicles can also be used for life estimation analysis. In such cases, usage history information is of key importance in order to correlate the test results with cells cycled in a controlled way – simply measuring the capacity will not be satisfactory to estimate or predict the remaining life of the battery.

Below, Li-ion cells will be used to illustrate the following analytical methods suitable for understanding degradation behaviour: *galvanostatic cycling analysis*, *impedance measurement*, *incremental capacity analysis* (ICA), *differential voltage analysis* (DVA), and *post-mortem analysis*.

7.3.1   Galvanostatic cycling

One of the most commonly used methods to observe capacity degradation is galvanostatic cycling. This method results in discharge profiles clearly showing the changes in the cell during ageing. Qualitative comparisons between cells at BOL and EOL, and for cells cycled a specific number of times or during a specific time period can be made.

In order to follow the degradation, this method is preferably used at low C-rates to enable observations of small changes in the discharge profiles, indicating both capacity fade and impedance rise. The discharge profiles can be studied as a function of SOC or capacity. If SOC is used, the discharge profiles are based on measured capacity; i.e. the reference capacity used to calculate the variations of SOC over time to allow 100% SOC to correspond to a fully charged cell regardless of capacity decline due to ageing. Discharge profiles for the two cases are illustrated in Figure 7.13.

In the SOC-based representations (Figure 7.13a), the impedance increase can be observed as higher voltage drops towards fully charged or discharged states, respectively, resulting mainly from slow mass transport within the active materials in the electrodes. Using capacity in the discharge profile representation (Figure 7.13b) may result in observations of changes in both cell characteristics and capacity decreases.

The galvanostatic cycling can be performed in different cycling conditions, the same or differing conditions during charge and discharge. The cycling conditions often refer

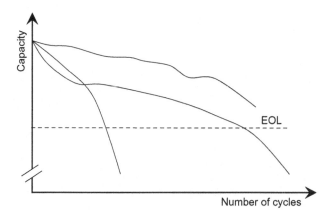

Figure 7.14 Normalised cell capacity as a function of energy throughput for cells cycled using different cycling conditions.

to the C-rates, based on symmetric cycles using, for example, 1C conditions, or asymmetric cycles with different C-rates during charging and discharging, all depending on the operation conditions of interest. When using a higher C-rate during the charge process, degradation under fast-charging conditions can be investigated.

The galvanostatic cycling can be made for long periods of time, all dependent on the desired outcome. Normally the cycling is interrupted and different cell performance characteristics, such as capacity or impedance (see below), are evaluated at predefined intervals, e.g. after a number of cycles have been performed or after a specific time. Normally, the tests continue until EOL conditions have been reached; often set to 80% of initial capacity (Section 5.2.2). The results obtained from galvanostatic cycling are normally related to the conditions at BOL (e.g. the capacity at BOL) and are presented as a function of time or energy throughput, as illustrated in Figure 7.14.

Results obtained from galvanostatic cycling are further used in the ICA and DVA methods to estimate cell/battery degradation.

7.3.2   Electrochemical impedance spectroscopy

Cells comprise of resistive, capacitive, and inductive components. The resistive features originate from the bulk electrode material, the electrolyte, the current collectors, and the charge-transfer reactions at the electrode/electrolyte interfaces. The double-layer capacitance is the main contributor to the capacitive part, and the inductive nature of the cell is mainly a result of the porosity of the electrodes. To use a specific cell parameter, which changes significantly with age, such as impedance, is one way to estimate degradation.

Often, electrochemical impedance spectroscopy measurements (EIS, Section 1.7.3) are performed – a standard, non-destructive technique of diagnosing the internal state of the cell and studying the ageing of cells – and a device can be incorporated in the battery installed in the vehicle. By applying a varying frequency signal, it is possible to

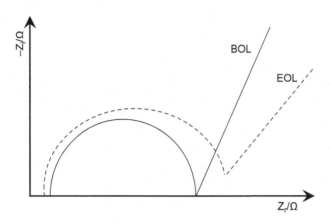

Figure 7.15 EIS spectra of a cell at BOL and at EOL, respectively.

distinguish between fast and slow processes occurring in the cell, and to obtain knowledge about several cell parameters. Slow processes are mainly related to mass-transport limited processes in the electrodes, as well as in the electrolyte, while interface reactions are often fast processes. In order to use EIS to study ageing effects, qualitative comparison of spectra is made. As illustrated in Figure 7.15, the ageing of the cells can be observed by larger semicircles shifted towards higher frequencies.

The semicircle arises from charge-transfer reactions at the interface between the electrolyte and the electrodes, and the tail reflects diffusion of lithium ions in the electrolyte and within the solid-phases of the active electrode materials. A wider semicircle refers to increased cell resistance, which is often related to irreversible side reactions and material degradation processes. The charge-transfer and the internal *IR* resistances increase over time, due to material degradation processes. During the degradation processes, the porosity of the electrodes may change significantly, which can be observed at high frequencies in the EIS spectrum as a result of the inductance changes in the cell.

Based on EIS measurements the remaining battery capacity, and corresponding life, can be further evaluated using equivalent circuit models. The data obtained by analysing commercial cells, often sealed, are not rarely a combination of parameters originating from both the positive and the negative electrodes and to separate the parameters corresponding to the individual electrodes involves assumptions. Direct estimation of the individual electrode parameters is only accessible using a reference electrode, impractical for cells incorporated in a battery for electric vehicles.

7.3.3   Incremental capacity

Most of the information regarding electrochemical and performance properties of the electrodes can be obtained from well-defined tests using a reference electrode (Section 1.7.4). The reference electrode will be held at a constant potential during the charge and discharge process, allowing one electrode at a time to be studied. This set-up is of interest

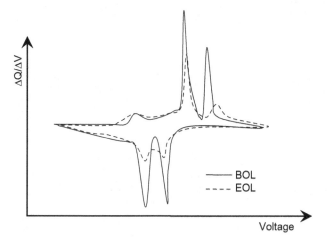

Figure 7.16 Schematic ICA profile; $\Delta Q/\Delta V$ as a function of cell voltage: BOL (full line), and EOL (broken line).

in order to study the loss of cyclable lithium and reduced capacity caused by loss of active electrode material. In most situations, however, a reference electrode cannot be employed, especially for cells used in the intended applications (e.g. a vehicle).

The *incremental capacity analysis* (ICA) is a method to identify and quantify changes in the electrochemical properties of the cell, relying on calculations of the differential capacity of small changes in the discharge profile and is expressed as $\Delta Q/\Delta V$. From a discharge profile obtained by galvanostatic cycling at slow rates, it is possible to analyse and quantify changes in the electrochemical properties of the cell. The method is used to identify parts in the discharge profile corresponding to the phase equilibrium of the active electrode materials; the flat plateaus in the discharge profile. At equilibrium, two or more phases with different lithium concentrations, but the same chemical potential coexist. Thus, the change in potential (ΔV) will be zero at the phase equilibrium and $\Delta Q/\Delta V$ is undefined. The method can be used for full cells and for half cells.

Based on changes in the $\Delta Q/\Delta V$ profiles, it is possible to identify ageing mechanisms, and one illustrative example is shown in Figure 7.16.

The width, amplitude, and distribution of the peaks appear at material-specific potentials and are strongly correlated to the changes in the active material during cycling of the cell. The separation between two peaks in the ICA profile should be constant for unchanged active electrode materials. In a Li-ion cell, the amount of lithium in an electrode varies and it is also a measure of SOC, and the potential varies accordingly. Therefore, it would be possible to evaluate the capacity decrease by exploring small and often negligible changes in the discharge profiles.

Due to material degradation, the peaks may be reduced in amplitude, wider, or shift. Symmetrical changes in the peaks indicate a symmetrical ageing of the cell, which would be the case if the active material of the positive and the negative electrodes were ageing at the same rate. Asymmetrical changes in the ICA profiles indicate other

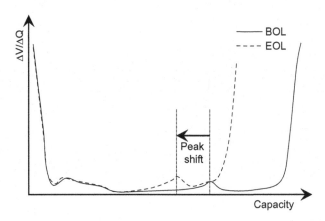

Figure 7.17 Schematic DVA profile; $\Delta V/\Delta Q$ as a function of cell capacity: BOL (full line), and EOL (broken line).

inhomogeneous ageing mechanisms. The shape and the intensity of the peaks provide information about the utilisation of the active electrode materials. A broadening of the peaks normally indicates slower electrode kinetics associated with transformations of the crystal structure or dissolution of the active electrode materials, i.e. loss of active electrode material reducing the overall cell capacity.

In an ICA profile, it is possible to detect different degradation mechanisms. The overall increase in cell impedance, or cell resistance, gives rise to increased voltage drops, which can be observed in the $\Delta Q/\Delta V$ profile as peak shifts. The peaks associated to the charge process (i.e. > 0 Ah/V) would shift to higher potentials and the peaks associated to the discharge process would shift to lower potentials (i.e. < 0 Ah/V). More detailed information will be obtained if half cells are used. In such cases, the peaks from the full cell measurements can be associated to specific changes in the active electrode materials and indicate film formation processes occurring at specific SOC values.

7.3.4   Differential voltage

The *differential voltage analysis* (DVA) method is closely related to the ICA method, and used to analyse the electrode contributions to the cell potential. Thereby the electrode utilisation as a function of cell age can be obtained. A DVA profile is basically the inverse of an ICA profile and is expressed as $\Delta V/\Delta Q$, and schematically illustrated in Figure 7.17.

The peaks in the DVA profile denote relatively fast voltage changes associated with phase transitions in the active electrode material, whereas the peaks in the ICA profiles represent phase equilibrium. As for the ICA method, the DVA profile must be obtained at low rates in order not to be affected by overpotentials caused by mass transport and charge transfer within the active electrode materials.

The DVA profile can be used to identify sources of capacity fade due to the electrodes either by side reactions or by loss of active material. A decrease in the peak distance in the DVA profile indicates a corresponding decrease in capacity. On the other hand, loss of cyclable lithium will not affect the peak distances. If capacity is lost as a result of loss of active electrode material from one of the electrodes, the corresponding peak will move relative to the other. If the loss of active electrode material is uniform, the peaks will both move. This is most likely due to an overall capacity change caused by imbalance of the electrodes. In addition, side reactions can give rise to new peaks in the DVA profile.

The DVA method will not identify the capacity fade mechanisms or where in the discharge profile the degradation initiated if the origin of the ageing mechanisms is required. Reliable DVA data can only be obtained from half cell measurements. Half cell measurements of both the positive and the negative electrodes can be used to identify phase transitions of the active electrode material. These appear as distinct peaks in the DVA profile and can be assigned to the peaks in the full cell profile. The position of these phase transition peaks relative to one another provides information about the source of the cell capacity fade and the type of fade responsible for the capacity degradation.

7.3.5   Half cell

As indicated above, reference electrodes are preferably used in order to identify the electrode responsible for the change in the cell performance and mostly affected by specific ageing conditions. Introducing a reference electrode in a real battery is, however, more or less impossible, even for a very limited number of cells per battery. Therefore, used cells are disassembled (more about this in Section 7.3.6) and half cells (Section 1.3.1) made out of the electrodes. This is a useful analysis method in order to correlate the observed degradations with a specific part of the cell, e.g. the positive or the negative electrode.

The half cell measurements will result in capacity and impedance data for the positive and the negative electrodes, and the method can be used for both fresh and aged cells. The drawbacks are the lack of understanding of full cell behaviour, and the fact that the electrodes are harvested from full cells. The half cell measurements give insight to the ageing mechanisms of full cells, but the influences of the electrodes on each other is lost.

Quite often laboratory half cells are used as the basis for the analysis of commercial cells. The positive and the negative electrodes are respectively cycled versus a metallic lithium foil as the counter electrode and the corresponding ICA and DVA profiles are obtained preferably using slow C-rates in order to capture the electrochemical processes.

Discharge profiles obtained from half cell measurements can be normalised according to the capacity of the electrode investigated and a full cell potential profile can be achieved as the difference between the two electrodes, as illustrated in Figure 7.18.

In order to match half cell profiles to a full cell profile, the half cell profiles have to be adjusted and scaled, indicating which of the electrodes limits cell capacity at both charged and discharged states. If one electrode is not fully utilised, the cell capacity is lower than the capacity of the limiting electrode. During ageing, one of the electrodes

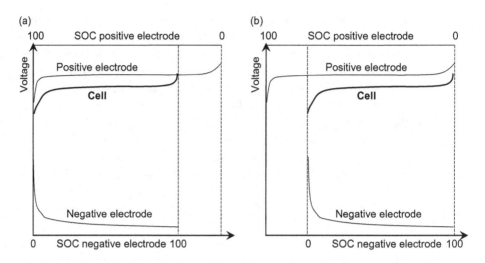

Figure 7.18 Differences in electrode balancing for a Li-ion cell having (a) excess of cyclable lithium, and (b) shortage of cyclable lithium.

may be utilised even more, while the other becomes less utilised. Increasing utilisation of one of the electrodes indicates loss of cyclable lithium.

The electrodes in Li-ion cells are balanced (Section 4.3.1) to ensure the potential of the negative electrode does not drop to zero, in order to prevent Li-plating (Section 3.1.3), as the cell is fully charged. Figure 7.18 illustrates the differences in the electrode balancing when there is an excess and a shortage of cyclable lithium. In Figure 7.18, the negative electrode limits the cell capacity. As the negative electrode is fully lithiated, i.e. the cell has 100% SOC, the positive electrode still has cyclable lithium available (Figure 7.18a). If there is a shortage of cyclable lithium (Figure 7.18b), the negative electrode is fully delithiated when still more lithium could be inserted into the positive electrode.

The discharge profile of the full cell is affected by several factors: e.g. the balance between the electrodes, capacity limitations of the individual electrodes, loss of cyclable lithium, gradual loss of active material during cycling. Cyclable lithium can also be lost directly from the electrolyte to be incorporated in the SEI layer as the cell is not being used, resulting in an imbalance even if both the electrodes preserve their respective lithium content, or SOC level. The lithium loss of the electrolyte will result in a loss of conductivity and thereby a loss of overall cell performance.

7.3.6   Post-mortem

Half cell measurements have also been shown to be important tools in *post-mortem analysis* in order to further understand the origin of degradation mechanisms. Performance decay and cell ageing originate from mechanical and chemical degradation of the composite electrodes and the electrolyte, both in terms of active and non-active materials. Many of the root-causes of degradation can be observed by opening used cells and gathering the harvested electrodes and, if possible, the electrolyte for analysis. This is a post-mortem analysis method since it is not applicable to cells in a battery during operation.

The opening of used cells for post-mortem analysis is not a trivial act. First of all, the cells must be opened in an inert atmosphere in order not to oxidise any reactive materials. The SOC of the cell must also be considered since opening a charged cell is constrained by safety considerations. Moreover, the opening must be performed with the utmost care due to the brittleness of the electrodes.

Parts of the electrodes can be used to assemble new cells for analysis. Half cells or full cells made of harvested electrodes may be applied for detailed investigation of degradation phenomena. Using half cells, it is possible to analyse, for example, the impedance of each electrode in order to understand how the degradation has proceeded for that particular electrode. The ICA and DVA methods described above are both applicable to these newly made cells. Capacity and impedance analysis, in combination with material analysis will provide deeper insights into the degradation mechanisms and help to assign the different observations to specific materials. If the cells are aged in a controlled way, i.e. in a laboratory, the post-mortem observations can more easily be correlated to the specific operational conditions, such as temperature and C-rates.

The post-mortem analyses are preferably carried out on electrode materials harvested at different positions within the cell in order to further understand the complexity of the degradation mechanisms. In the case of cylindrical cells, the ageing conditions may vary between the inner part of the coil and the outer part, and for pouch cells there may be differences between the parts near the current tabs and the parts further away.

Additional understanding of the cell degradation mechanisms can be obtained by analysing the electrolyte, which may give information about, for example, salt degradation and electrode material dissolutions into the electrolyte.

Glossary

Active material The material active in the electrochemical *redox* reactions at either the *positive* or the *negative electrode*.

Additive A substance added in small amounts to the electrolyte to enhance performance.

Ageing A permanent loss of performance.

Ampere (A) The SI unit for electric *current*; named after André-Marie Ampère (1775–1836).

Ampere hour (Ah) The unit of *electric charge*. 1 Ah is equal to 3600 *coulombs*.

Anode The *electrode* at which the *oxidation* takes place, i.e. the electrode where the electrons are ejected.

Available capacity The fraction of the total *capacity* that can be used at defined charge and discharge conditions.

Battery A group of series-and/or parallel-connected *cells*, including support functions, control unit, and thermal management, intended for a specific application.

Bipolar electrode An electrode functioning as *anode* of one *cell* and *cathode* of another *cell*.

C-rate A measure of the (dis)charge rate relative to the maximum *cell capacity*. lC rate is a discharge current discharging the cell within 1 hour.

Calendar life The elapsed time before a battery becomes unusable whether it is in active use or inactive.

Capacitance The ability of a *capacitor* to store *electrical charge*. The SI unit is *farad*.

Capacitor A capacitor stores energy in an electric field and contains at least two electrical conductors separated by a dielectric (insulator).

Capacity The amount of *electric charge* that can be stored. Normally expressed in Ah.

Capacity retention The fraction of the total *available capacity* usable after storage.

Cathode The *electrode* at which the *reduction* takes place, i.e. the electrode where the electrons are accepted.

Cell See *electrochemical cell*.

Charge See *electric charge*.

Charge carrier An ion or electron free to move, carrying the ionic or *electric charge*, respectively.

Charge current The current used during charging of a *cell*.

Charge current limit Maximum *charge current* allowed, restricted to avoid cell damage.

Constant current charging A charging method using a constant *current*, allowing the *voltage* to vary.

Constant voltage charging A charging method applying a fixed *voltage*, allowing the *current* to vary.

Coulomb (C) The derived unit of *electric charge*, defined as the charge transferred by a steady current of one ampere for one second. One coulomb is also the amount of excess charge on the positive side of a *capacitance* of one *farad* charged to a *potential difference* of one *volt*. Named after Charles-Augustin de Coulomb (1736–1806).

Coulombic efficiency The ratio of the *electric charge* introduced to the *cell* during charging versus the *electric charge* extracted from the *cell* during discharging.

Current The flow of *electric charge*. The SI unit for measuring an electric current is *ampere*.

Cut-off voltage Voltage above/below which a *cell* should not be charged/discharged.

Cycle life The number of charge and discharge cycles a *cell* can experience before the *capacity* fails to meet specific performance criteria.

Cycling stability Number of cycles which can be achieved within a specified capacity range.

Depth of discharge (DOD) The *capacity* that has been discharged from a *cell* expressed as a percentage of maximum capacity.

Discharge current limit Maximum *discharge current* allowed, restricted to avoid cell damage.

Efficiency See *Coulombic efficiency*.

Electric charge The basic property of matter carried by some elementary particles. Electric charge, which can be positive or negative, occurs in discrete natural units and is neither created nor destroyed.

Electrical equivalent model An electrical circuit approximating the behaviour of the *electrochemical cell*.

Electrochemical cell Consists of two *electrodes* separated by an *electrolyte*, capable of either deriving electrical energy from chemical reactions, or facilitating chemical reactions through the introduction of external electrical energy.

Electrochemical potential The work done in bringing one mole of an ion from a standard state to a specified electrical potential. Electrochemical potential is expressed in the unit of J/mol.

Electrode An electrical conductor used to make contact with a non-metallic part of a circuit (e.g. a semiconductor, an electrolyte, or a vacuum).

Electrode potential The potential obtained by an *electrode* against a standard reference electrode.

Electrolyte An ion-conducting media separating the *electrodes*. Normally consists of a *salt* dissolved in a solvent.

Energy density The amount of *energy* stored per unit volume.

Farads (F) The SI unit of *capacitance*, named after Michael Faraday (1791–1867). A *capacitor* of 1 F charged with 1 *coulomb* of *electrical charge* will have the *potential difference* of 1 *volt* between its plates.

Fuel cell A *galvanic cell* externally supplied by *active materials*.

Galvanic cell An *electrochemical cell* converting chemical energy to electric energy by electrochemical reactions. Named after Luigi Galvani (1737–1789).

Half cell A cell of one *electrode* immersed in *electrolyte* measured against a *reference electrode*.

Impedance An expression of *resistance* and reactance according to how a cell retroacts to alternating and/or direct *current*.

Internal resistance The *resistance* to the flow of electrical *current* within the *electrochemical cell*.

Internal short circuit An electrical circuit allowing *current* to pass along an unintended path inside the *electrochemical cell*.

IR-drop A *voltage* drop in a *cell* related to the *internal resistance* and *current* used.

Negative electrode During discharge, the *anode* is the negative electrode. During charge, the *cathode* is the negative electrode.

Nominal voltage The *voltage* used to express the *rated capacity*.

Open circuit voltage (OCV) The difference of *electrochemical potential* between two *electrodes* of a device when disconnected from any external circuit.

Overcharge Charging of an *electrochemical cell* continued after a fully charged state is reached, i.e. a cell charged above the upper *cut-off voltage*.

Overdischarge Discharging of an *electrochemical cell* beyond a fully discharged state is reached, i.e. a cell discharged below the lower *cut-off voltage*.

Overpotential The difference between an equilibrium *potential* and the *potential* at which the *redox* reactions are experimentally observed.

Oxidant The species being oxidised in a *redox* reaction.

Oxidation The loss of electrons or an increase in oxidation state of a molecule, atom, or ion.

Polarisation The *potential* change of a *cell* or *electrode* from its equilibrium state caused by an applied *current*.

Polarity reversal Reversing of polarity of the *electrodes* due to *overdischarge*.

Positive electrode During charging, the *anode* is the positive electrode. During discharge, the *cathode* is the positive electrode.

Potential See *electrochemical potential*.

Power The rate at which electric *energy* is transferred by an electric circuit. The SI unit of power is the watt (W) (=1 J/s).

Power density The amount of power extractable per unit volume.

Primary cell An *electrochemical cell* in which the *redox* reaction cannot be reversed; the *cell* is not able to be recharged.

Rate capability Describes highest possible *current* without damaging the cell.

Redox Short for *reduction-oxidation*. All chemical reactions in which atoms/ions undergo changes in oxidation state.

Reductant The species reduced in a *redox* reaction.

Reduction The gain of electrons or a decrease in oxidation state of a molecule, atom, or ion.

Reference electrode An *electrode* having a stable and well-known *electrode potential*, against which other electrode potentials are normalised.

Salt The source of *charge carriers* in the *electrolyte*.

Secondary cell An *electrochemical cell* in which the chemical reactions are reversible, i.e. it can be recharged.

Self-discharge Internal chemical reactions reducing the stored charge of the *electrochemical cell* without any connection between the *electrodes*.

Shelf-life The time a *cell* or *battery* can be stored with remaining acceptable performance.

Specific energy The amount of *energy* stored per unit mass.

Specific power The amount of *power* extractable per unit mass.

State of charge (SOC) The *capacity* as a percentage of maximum capacity, a measure of the amount of *charge* stored in the *cell*.

State of health (SOH) The total remaining *capacity* of a *cell* compared to a fresh *cell*.

Undercharge See *overdischarge*.

Volt (V) The SI unit for *electrochemical potential*, named after Alessandro Volta (1745–1827).

Voltage The *electrochemical potential* difference between the *electrodes* with load applied. Equal to the electromotive force of the cell.

Working voltage The actual operating *voltage* for an *electrochemical cell* under load. Always lower than the *OCV*.

Further reading

R.A. Huggins, *Advanced batteries*, Springer, New York, 2009.

J.O. Besenhard (ed.), *Handbook of Battery Materials*, Wiley-VCH Verlag GmbH & Co. KGaA, Weinheim (Germany), 1999.

C.M.A. Brett and A.M.O Brett, *Electrochemistry: Principles, Methods, and Applications*, Oxford University Press, Oxford, 1993.

J. Garche (ed.-in-chief), *Encyclopedia of Electrochemical Power Sources*, Elsevier, Amsterdam, The Netherlands, 2009.

J.-K. Park, *Principles and Applications of Lithium Secondary Batteries*, Wiley-VCH Verlag GmbH & Co. KGaA, New York, 2012.

T.B. Reddy and D. Linden, *Handbook of Batteries*, 4th edn, McGraw-Hill, New York, 2011.

M. Yoshio, R.J. Brodd, and A. Kozawa (eds.), *Lithium-Ion Batteries*, Springer, New York, 2009.

Index

18650 cell, 134
3D cell design, 13

A_2B_7, 56
AB_2, 56
AB_5, 56
absorbed glass mat battery, 52
activated carbon, 65
activation polarisation, 33
  fuel cell, 75
active battery balancing, 186
active cooling, 164
active electrode area, 63
active material, 8, 230
additive, 230
additives, 115, 118
  film-forming, 120
  flame-retardant, 120
  gas-forming, 120
  redox-shuttle, 119
  SEI forming, 119
  shut-down, 120
ageing, 230
ageing rate, 189
AGM, 52
air-cooling, 164
Al current collector, 128
all-electric vehicle, 144
alloy
  negative electrode, 96
amorphous carbon, 91, 95, 127
ampere, 230
ampere hour, 230
anode, 7, 230
anodic current, 22
assembly process, 138
asymmetrical capacitor, 65
available capacity, 230

β-Al_2O_3, 67
band structure, 31
battery, 14, 230
battery balancing, 185

battery degradation, 194
  accelerating factors, 201
  active materials and electrolytes, 212
  analysis methods, 221
  changes in bulk material, 210
  current, 206
  electrolytes, 219
  layered oxides, 217
  $LiFePO_4$, 219
  $LiMn_2O_4$, 218
  mechanisms, 195
  negative electrode, 212
  positive electrode, 215
  SEI, 213
  separators, 220
  temperature, 202
  voltage, 206
battery design, 152
battery durability, 194
battery electric vehicle, 144
battery life, 155, 189
battery management system, 168
battery models, 183
  cycle counting, 191
  empirical, 191
  energy throughput counting, 191
  equivalent circuit, 191
  physical, 191
battery monitoring, 178
beginning-of-life, 155
BEV, 144
binder, 127
bipolar cell, 12
bipolar electrode, 230
bipolar plate, 76
blocking electrode, 9
BMS, 168
BOL, 155
Butler–Volmer equation, 22

calendar ageing, 195
calendar life, 155, 189, 195, 230
calendering, 138

capacitance, 63, 230
capacitor, 62, 230
capacity, 36, 230
    theoretical, 37
capacity fade, 195
capacity retention, 230
carbon black, 93, 127
cathode, 7, 230
cathodic current, 22
cell, 230
cell balancing, 130
cell casing, 11
cell design, 126, 134
cell format, 134
cell formation, 139
cell production, 137
cell selection, 156
cell voltage, 17
cell-to-cell variations, 157
centralised BMS, 171
charge, 230
charge carrier, 230
charge current, 230
charge current limit, 230
charge separation distance, 63
charge transfer, 22
charge-depleting, 150
charge-sustaining, 150
charge-transfer resistance, 44
charging methods, 171
    constant current, 172
    constant current–constant voltage, 173
    constant voltage, 173
    temperature control, 172
    voltage control, 172
chemical potential, 16
Chevrel phases, 72
CID, 137
composite electrode, 28, 126, 129
concentration polarisation, 25, 33
    fuel cell, 75
constant current charging, 231
constant power discharge rate, 36
constant voltage charging, 231
constant voltage charging method, 173
contaminations, 211
conversion material, 30
conversion oxides, 98
conversion reaction, 30
cooling media, 164
cooling system, 163
core-shell model, 110
co-solvent, 116, 125
coulomb, 231
coulombic efficiency, 40, 231
counter electrode, 46

C-rate, 35, 230
crossover, 78
Cu current collector, 128
current, 22, 231
current collector, 10, 126
    corrosion, 214
current counting, 182
current density
    fuel cell, 75
current distribution, 128
current interrupt device, 137
cut-off voltage, 21, 35, 231
cycle life, 155, 189, 231
cycling stability, 231
cylindrical cells, 134

degradation mechanisms
    chemical, 195
    Li-ion cell, 208
    mechanical, 195
    side reactions, 210
    surface film formation, 210
degrees of electrification, 144
dendrites, 90
depth of discharge, 181, 231
desulphation, 50
dielectric constant, 63
differential voltage analysis, 226
diffusion coefficient, 25
diffusion pathway, 29
dimethyl carbonate, 116
discharge current limit, 231
discharge profile, 17
DMC, 116
DOD, 181, 231
double-layer capacitance, 24, 44, 63
double-layer capacitor, 65
drying process, 138
duty cycle, 147

EC, 116
EDLC, 65
efficiency, 231
electric charge, 231
electrical equivalent model, 231
electrical impedance, 44
electrical potential, 16
electrochemical cell, 7–8, 231
electrochemical double-layer, 24
electrochemical double-layer capacitor, 62
electrochemical equilibrium, 16
electrochemical force, 15
electrochemical impedance spectroscopy, 44, 223
electrochemical potential, 16, 231
electrochemical stability window, 32
electrode, 7, 9, 77, 231

electrode fabrication, 137
electrode kinetics, 22
electrode potential, 231
electrolyte, 7, 9, 231
  concentration, 25
  gel polymer, 123
  ionic liquids, 124
  Li-ion battery, 114
  liquid, 115
  polymer, 123
  resistance, 44
  viscosity, 26
electrolytic cell, 7
electronic protection devices, 165
electronic structure, 30
electrostatic field, 62
electrostatic potential, 16
end-of-life, 155
energy, 38
  of capacitor, 62
energy density, 38, 231
energy efficiency, 41
energy management, 152
energy optimised cell, 130
energy optimised electrode, 129
energy throughput, 147
entropy heat, 21
EOL, 155
ethylene carbonate, 116
EV, 144
exchange current, 22
exfoliation, 95, 214

farad, 231
Faraday's law, 37
fast-charging, 155
fault detection, 170
Fermi level, 30
Fick's first law, 25
Fick's second law, 26
film-forming additives, 120
flame-retardant additives, 120
fluorophosphates, 112
fluorosulphates, 112
forced air-cooling, 164
formation reaction, 30
fuel cell, 74, 232
full cell, 12

Galvani potential difference, 16
galvanic cell, 7, 232
galvanostatic cycling, 41, 222
gas diffusion layer, 77
gas formation additives, 120
gas-shift reaction, 79
gel battery, 52

gel polymer electrolyte, 123–4
Gibbs free energy, 15–16, 21
Gibbs phase rule, 17
graphite, 91

half-cell, 11, 227, 232
hard carbons, 96
heating media, 164
heating system, 163
Helmholtz plane, 24
HEV, 145
high-temperature battery, 66
HOMO, 31
hot-spots, 177
hourly rate, 36
Hunter's disproportional reaction, 108
hybrid capacitor, 65
hybrid electric vehicle, 145
hydrogen, 74

impedance, 232
incremental capacity analysis, 225
inner Helmholtz plane, 24
inorganic composite membrane separator, 122
insertion
  negative electrode, 98
insertion electrode, 9
insertion material, 29
intercalation, 29
intermetallic alloy, 55
intermetallic materials
  negative electrode, 97
internal resistance, 33, 232
  fuel cell, 75
internal short circuit, 176, 232
ion conductivity, 26–7
ion mobility, 26
ion pairing, 27
ionic liquids, 66, 124
IR drop, 33, 232
irreversible heat generation, 21

Jahn–Teller distortion, 107, 218
Joule heating, 21

KOH, 56

layered oxides, 102
lead dioxide, 48
lead–acid battery, 48
LFP, 109
Li (metal) polymer battery, 83
Li metal battery, 83
Li_2FeSiO_4, 112
$Li_2Mn_2O_4$, 107
Li_2MnO_3, 105

Li_2MSiO_4, 112
$Li_4Ti_5O_{12}$, 98
Li-air battery, 59
$LiAsF_6$, 117
$LiBF_4$, 117
$LiCF_3SO_3$, 117
$LiClO_4$, 117
$LiCoO_2$, 102
$LiCoPO_4$, 111
Li-dendrites, 205
$LiFePO_4$, 109
Li-ion battery, 59, 83
Li-ion capacitor, 65
Li-ion polymer battery, 59, 83
$LiMn_2O_4$, 105
$LiMnPO_4$, 111
$LiMPO_4F$, 112
$LiMSO_4F$, 112
$LiN(SO_2CF_3)_2$, 117
$LiNi_{0.8}Co_{0.15}Al_{0.05}O_2$, 104
$LiNi_{1/3}Mn_{1/3}Co_{1/3}O_2$, 104
$LiNiO_2$, 103
Li-oxygen battery, 59
$LiPF_6$, 117, 220
Li-plating, 93, 205, 211
liquid cooling, 164
liquid electrolyte, 115
Li-rich NMC
    Li-rich, 105
Li-Si alloys, 97
Li-sulphur battery, 60
LiTFSI, 117
lithium, 57
lithium batteries, 57
lithium metal battery, 58
lithium-ion battery, 83
Li-triflate, 117
$LiVPO_4F$, 113
$Li_xM_2(XO_4)_3$, 111
LTO, 98
LUMO, 31

many-particle model, 110
mass transport, 24, 28
master–slave BMS, 171
MEA, 77
mechanical stability, 154
Me-ion battery, 70
membrane
    fuel cell, 74
membrane electrode assembly, 77
memory effect, 57
metallic lithium
    negative electrode, 89
Mg battery, 72
micro-HEV, 145
microporous polymeric membrane separator, 122

mild HEV, 145
modular BMS, 171
molten salts, 67
monopolar cell, 12
multi-layered separator, 122

N/P ratio, 55, 132
Na-ion battery, 70
Na-NiCl$_2$ battery, 67
natural gas, 74
NCA, 104
negative electrode, 7, 232
    Li-ion battery, 86
Nernst equation, 17, 49
nickel metal-hydride battery, 52
nickel oxyhydroxide, 53
NiMH, 52
NiOOH, 53, 56
NiZn battery, 68
NMC, 104
nominal voltage, 17, 232
non-active material, 8
non-aqueous solvents, 115
non-blocking electrode, 26
non-woven fabric mat separator, 122
Nyquist plot, 44

OCV, 17, 33, 232
ohmic polarisation, 33
ohmic resistance
    fuel cell, 75
open circuit, 44
open circuit voltage, 17, 232
operating temperature range, 202
outer Helmholtz plane, 24
overcharge, 35, 232
overdischarge, 35, 232
overpotential, 22, 232
oxidant, 7, 232
oxidation, 7, 232
oxides
    negative electrode, 98
oxygen recombination reaction, 54

P/E ratio, 148
parallel connected capacitors, 64
passive battery balancing, 186
passive cooling, 164
Pb-acid, 48
PbO_2, 48
PC, 116
PEM, 77
PEMFC, 75
PEO, 123
Peukert law, 37
Peukert plot
    modified, 37

PEV, 144
PHEV, 146
planar electrode, 28
plasticiser, 123
plug-in hybrid electric vehicle, 146
polarisation, 232
polarisation profile, 75
polarity reversal, 55, 232
poly(ethylene oxide), 123
poly(propylene oxide), 123
polyanionic materials, 111
polymer electrolyte, 123
polymer electrolyte membrane, 77
polymer electrolyte membrane fuel cell, 75
polysulphides, 61
porous electrode, 28
positive electrode, 7, 232
    Li-ion battery, 100
    mixed materials, 113
positive temperature coefficient, 137
post-mortem analysis, 228
potassium hydroxide, 56
potential, 232
potential difference, 15
potentiostatic cycling, 41
pouch cells, 135
power, 39, 232
power capability, 22
power density, 232
power fade, 195
power optimised cell, 130
power optimised electrode, 129
power-to-energy ratio, 148
PPO, 123
primary cell, 7, 232
prismatic cells, 135
propylene carbonate, 116
pseudo-capacitance, 64
pseudo-capacitor, 65
PTC, 137
pure electric vehicle, 144

Ragone plot, 39, 132
rate capability, 232
rated cell capacity, 35
redox flow battery, 72
redox reaction, 7, 232
redox-shuttle additives, 119
reductant, 7, 232
reduction, 7, 232
reference electrode, 12, 46, 233
reference potential, 17

salt, 233
    Li-ion battery, 117
salt concentration, 27, 118
secondary cell, 7, 233

SEI, 87
SEI forming additives, 119
self-discharge, 155, 198, 233
    Pb-acid, 50
separator, 10, 120
    inorganic composite membrane, 122
    microporous polymeric membrane, 122
    multi-layered, 122
    non-woven fabric mat, 122
    permeability, 121
    pore size, 121
    shut-down, 121
series connected capacitors, 64
shedding, 51
shelf-life, 155, 233
shut-down additives, 120
Si-based electrode, 96
silicon, 96
single-phase insertion reaction, 30
SLI, 51
SOC, 180, 233
SOC estimation
    capacitors, 187
SOF, 192
SOFC, 79
SOH, 189, 233
solid electrolyte interphase, 87
solid oxide fuel cell, 79
solid polymer electrolyte, 123
solvation shell, 95
solvent co-insertion, 214
solvents, 115
SOP, 192
specific capacity, 37
specific energy, 38, 233
specific power, 232
spinel, 105
    high-voltage, 108
standard hydrogen electrode, 17
standard potential, 15, 17
starting, lighting and ignition battery, 51
state functions, 179
state of charge, 180, 233
state of function, 192
state of health, 189, 233
state of power, 192
Stokes–Einstein equation, 26
stratification, 50
strong HEV, 145
sulphation, 50
sulphuric acid, 48
super capacitor, 62
surface protective layer, 101

tab, 136
tap density, 137
technology selection, 157

temperature, 20
temperature control charging method, 172
thermal control, 174
thermal cycling, 205
thermal management, 174
thermal runaway, 175
thermal system, 163
transfer coefficient, 22
transference number, 26
turbostratic disordered graphite, 93
two-phase insertion reaction, 30

ultra capacitor, 62
undercharge, 233
usage condition, 146

valve regulated battery, 52
vanadium redox flow battery, 73
VC, 119

vinylene carbonate, 119
Volt, 233
voltage, 15, 233
voltage control charging method,
    172
voltage depression, 57
voltage hysteresis, 34
voltage look-up table, 183
voltage profile, 17
voltammogram, 42
volumetric capacity, 37
VRLA, 52

Warburg impedance, 45
working electrode, 46
working voltage, 233

ZEBRA battery, 67
Zinc-air battery, 69

Printed in the United States
by Baker & Taylor Publisher Services